BIOLOGICAL SEQUENCE
ANALYSIS USING THE
SEQAN C++ LIBRARY

CHAPMAN & HALL/CRC
Mathematical and Computational Biology Series

Aims and scope:

This series aims to capture new developments and summarize what is known over the whole spectrum of mathematical and computational biology and medicine. It seeks to encourage the integration of mathematical, statistical and computational methods into biology by publishing a broad range of textbooks, reference works and handbooks. The titles included in the series are meant to appeal to students, researchers and professionals in the mathematical, statistical and computational sciences, fundamental biology and bioengineering, as well as interdisciplinary researchers involved in the field. The inclusion of concrete examples and applications, and programming techniques and examples, is highly encouraged.

Series Editors

Alison M. Etheridge
Department of Statistics
University of Oxford

Louis J. Gross
Department of Ecology and Evolutionary Biology
University of Tennessee

Suzanne Lenhart
Department of Mathematics
University of Tennessee

Philip K. Maini
Mathematical Institute
University of Oxford

Shoba Ranganathan
Research Institute of Biotechnology
Macquarie University

Hershel M. Safer
Weizmann Institute of Science
Bioinformatics & Bio Computing

Eberhard O. Voit
The Wallace H. Couter Department of Biomedical Engineering
Georgia Tech and Emory University

Proposals for the series should be submitted to one of the series editors above or directly to:
CRC Press, Taylor & Francis Group
4th, Floor, Albert House
1-4 Singer Street
London EC2A 4BQ
UK

Published Titles

Chapman & Hall/CRC Mathematical and Computational Biology Series

BIOLOGICAL SEQUENCE ANALYSIS USING THE SEQAN C++ LIBRARY

ANDREAS GOGOL-DÖRING
KNUT REINERT

CRC Press
Taylor & Francis Group
Boca Raton London New York

CRC Press is an imprint of the
Taylor & Francis Group an **informa** business
A CHAPMAN & HALL BOOK

CRC Press
Taylor & Francis Group
6000 Broken Sound Parkway NW, Suite 300
Boca Raton, FL 33487-2742

First issued in paperback 2017

© 2010 by Taylor and Francis Group, LLC
CRC Press is an imprint of Taylor & Francis Group, an Informa business

No claim to original U.S. Government works

ISBN 13: 978-1-138-11282-7 (pbk)
ISBN 13: 978-1-4200-7623-3 (hbk)

Library of Congress Cataloging-in-Publication Data

Gogol-Döring, Andreas.
    Biological sequence analysis using the SeqAn C++ library / Andreas Gogol-Döring, Knut Reinert.
        p. cm. -- (Mathematical and computational biology)
    Includes bibliographical references and index.
    ISBN 978-1-4200-7623-3 (hardcover : alk. paper)
    1. Information storage and retrieval systems--Nucleotide sequence. 2. Nucleotide sequence--Data processing. I. Reinert, Knut. II. Title.

QP625.N89G64 2009
611'.0181663--dc22   2009036644

Visit the Taylor & Francis Web site at
http://www.taylorandfrancis.com

and the CRC Press Web site at
http://www.crcpress.com

to Birgit, Kyra, and Finley

and

to Fabienne

Contents

Preface

The History of SeqAn

Before setting up the Algorithmic Bioinformatics group at the
Freie Universität Berlin, Knut Reinert had been working for years
at a U.S. company – Celera Genomics in Maryland – where he
worked on large genome assembly projects. A central part of these
projects was the development of large software packages containing assembly and analysis algorithms developed by the Informatics Research team at Celera. Although successful, the endeavor
clearly showed the lack of available implementations in sequence
analysis, even for standard tasks. Implementations of needed algorithmic components were either not available, or hard to access
in third-party, monolithic software products.

This experience in mind, and being educated at the Max-Planck-
Institute for Computer Science in Saarbrücken – the home of very
successful software libraries like LEDA (Mehlhorn and Näher 1999)
and CGAL (Fabri, Giezeman, Kettner, Schirra, and Schönherr
2000) – Knut set the development of such a software library high
on his research agenda.

The fundamental idea was that the library should be comprehensive for the field of sequence analysis, it should be easy to use,
and most of all (having the tremendous data volumes in genomics
in mind) be efficiently implemented. In 2003, Andreas Gogol-
Döring joined the Algorithmic Bioinformatics group. In the next
18 months, lively discussions about goals, different software designs, and the possible content of the library followed which led to
various prototypes that allowed us to verify the design ideas with

the corresponding implementations. Although this approach was rather work-intensive, it led to a lot of insights and finally to the current SeqAn design which in our opinion fulfills our initial goals. In this first phase Andreas bore the main implementation work. He was aided by some B.Sc. and M.Sc. students of the Bioinformatics curriculum at the FU Berlin. In this mode of operation, even with hard work, SeqAn would not have become what it is today. For this to happen, we were very lucky to be able to attract a handful of very talented Ph.D. students who joined the project. In 2006, David Weese and Tobias Rausch joined the SeqAn team followed by Anne-Katrin Emde in 2008. Their help in augmenting the functionality of SeqAn and in implementing algorithms, data types, and providing documentation and tutorials was indispensable in making SeqAn a great product.

SeqAn is currently used by several leading companies in the field as well as various research groups in Europe and the USA. The development is secured in the future: SeqAn will be improved, expanded and soon support multicore platforms.

The SeqAn Team

At the time of the writing of this book, the following people formed the main SeqAn team and contributed significant parts of the library, its documentation, and the supporting infrastructure.

Andreas Gogol-Döring started the SeqAn project together with Knut Reinert in the spring of 2003. He designed SeqAn and participated in basically every major project undertaken with SeqAn.

Knut Reinert initiated the SeqAn project and started working on it together with Andreas Gogol-Döring in the spring of 2003. He is the overall coordinator for staffing and research directions in SeqAn.

Anne-Katrin Emde joined the SeqAn project in the spring of 2008 and is the latest addition to the team. She implemented the multiple segment match refinement.

Tobias Rausch joined the team in October 2006. He is the main contributor to the graph type in SeqAn. Apart from standard algorithms on graphs and automata, he implemented the graph-based, multiple alignment strategy in SeqAn.

David Weese joined in 2006. He is the main contributor to the string indices and mapping tools in SeqAn.

In addition to the core SeqAn team, many other people contributed to the book, in particular, Tobias Marschall, Marcel Martin, and Sven Rahmann, who wrote Chapter 16, and Simona Rombo, Filippo Utro and Raffaele Giancarlo, who wrote Chapter 15. And last but not least, many external Ph.D. and M.Sc. students made valuable contributions to SeqAn. Our thanks go to Ji-Hyun Lim, Carsten Kemena, Marcel Schulz, and others.

About this Book

Part I gives an introduction to the SeqAn project, and it describes the general *library design*. Chapter 1 introduces the reader briefly to biological sequence analysis problems and the benefit of using software libraries. In Chapter 2 we summarize the general goals and design principles of SeqAn. More details follow in Chapter 3 where we state the main goals we want to achieve with the library design. The means to reach these goals are proposed in Chapter 4, where the main programming techniques used in SeqAn are elaborated on. The application of these techniques is demonstrated using examples in Chapter 5.

Part II explicitly describes the components provided by SeqAn. After proposing some basic functionality in Chapter 6, we describe sequence data structures in Chapter 7, alignments in Chapter 8, pattern and motif searching in Chapters 9 and 10, string indices in Chapter 11, and graphs in Chapter 12.

Part III presents some applications of SeqAn. In Chapter 13 we use a well-known genome alignment program – LAGAN – to illustrate how complex analyses can be quite easily programmed in SeqAn. We show how LAGAN can be implemented in about 200 lines of code without losing any efficiency. In Chapter 14 we demonstrate the versatility of the multiple sequence alignment component in SeqAn. It can be configured to serve a multitude of alignment tasks ranging from protein alignment to the computation of a consensus sequence in assembly projects. The last two chapters (15, 16) address the algorithm engineers. Chapter 15 shows how to add new functionality to SeqAn using the algorithmic components already present in the library and Chapter 16 gives a very nice example of how to incorporate a new algorithm (in this case for the construction of a suffix array) into SeqAn.

Andreas Gogol-Döring and Knut Reinert

Part I

The SeqAn Project

In Part I, we first discuss the role of *sequence analysis* in the life sciences, and explain how *software libraries* could facilitate the development of new software tools and algorithms for sequence analysis. SeqAn is the only software library available that focuses explicitly on the development of *highly performant* sequence analysis software by providing a comprehensive collection of the common algorithmic components and data structures. Chapter 2 gives a short overview of the SeqAn project and our measures for quality assurance and dissemination of the library.

SeqAn relies on a unique generic *design*. We explain the main goals that we pursued by the library (Chapter 3), as well as the programming techniques that we applied to achieve these goals (Chapter 4). These techniques are illustrated by some examples in Chapter 5.

Chapter 1

Background

1.1 Sequences in Bioinformatics

Sequences play a major role in biology as a means of abstraction. For example *deoxyribonucleic acid* (DNA), the carrier of genetic information in the nucleus, as well as *proteins*, a main ingredient of the cell responsible for most biological activity, can be represented as sequences over an alphabet of four, respectively twenty characters. This is due to the fact that those molecules are *biopolymers*, large organic molecules assembled from small building blocks called *monomers*, which are all of the same kind and linked together to long chains. The monomers of nucleic acids like DNA or RNA (*ribonucleic acid*) are *nucleotides*, and each nucleotide contains one out of four possible *nucleobases*. The structure of a nucleic acid strand is therefore defined by the actual sequence of bases in its nucleotides. Proteins on the other hand are composed of *amino acids*. In natural proteins, twenty different kinds of amino acids occur. They all have a phosphate backbone and differ in their *residues*. In proteins, these amino acids may occur in any order and number. We call the information about the succession of the monomers in a nucleic acid and protein its *biological sequence*, and thus we consider these biopolymers a kind of *storage* for this information. Many functions which are fulfilled by biopolymers like nucleic acids and proteins depend on their sequence composition. A DNA sequence for example encodes *genes*, which are construction plans for proteins. The cell first *transcribes* the genes into messenger RNA (*mRNA*), which is then, after some modifications, *translated* into a peptide, where every three nucleobases form a *codon* that corresponds to one specific amino acid in the synthesized protein. The sequence of nucleotides in the DNA therefore defines the order of amino acids in the protein, which further specifies the three-dimensional shape the protein folds into. RNA may also fold into a structure that is crucial to fulfill its purposes in the cell. Moreover the degree of molecular binding between proteins and nucleic acids depends on their sequences; the protein synthesis for example involves certain proteins that can *dock* only on specific patterns in the DNA.

A deeper understanding of biological processes thus requires a broad knowledge of the biomolecule sequences, and in the last decades a lot of research was aimed to decode those sequences. The most prominent projects in this field were certainly the *Human Genome Project* (International Human Genome Sequencing Consortium 2001) and its counterpart by Celera Genomics (Venter et al. 2001) which both aimed to sequence the entire human genome. Decrypted biological sequences are deposited in public databases as *strings*, i.e., ordered sequences of characters from a finite alphabet Σ. The succession of bases in a DNA can for example be stored in a string of the alphabet $\Sigma = \{A, C, G, T\}$, where each letter stands for one nucleic base, e.g., *A* for *adenine*. Figure 1 shows the rapid progression of the data volume deposited in Gen-Bank (Benson et al. 2008) and UniProtKB/Swiss-Prot (UniProt Consortium 2008). The number of nucleotides stored in GenBank has doubled approximately every 20 months and thus risen in two decades by four orders of magnitude. The protein database Swiss-Prot grew somewhat slower: From the beginning of the 1990s, the amount of amino acids increased about 20% per year, i.e., it doubles every four years.

Lately, several new sequencing technologies like *pyrosequencing* (also known as Roche/454 sequencing; Margulies et al. 2005) or *sequencing-by-synthesis* (also known as Illumina/Solexa technology; Bentley 2006) were invented, and they allow a much higher throughput than previous approaches. Hence, the size of the databases is expected to grow even faster in the future, since the availability and decreasing cost of sequencing open the door to new applications in metagenomics or personalized medicine. The analysis of these data may help to explain processes in the cell,

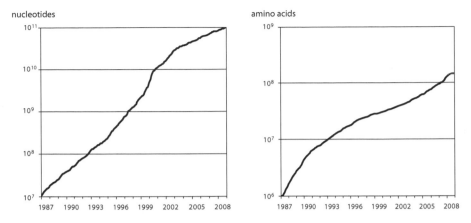

Figure 1: **Sequence Database Growth. Left**: Total number of bases stored in GenBank from the beginning of the year 1987 to the end of 2008. **Right**: Total number of million residues that were stored in UniProtKB/Swiss-Prot database during the same period. The data were taken from the release notes. Note the logarithmic scale.

like the regulation of gene expression, and this understanding offers the opportunity to develop new treatments of diseases or improved agricultural crops, and it will give us deeper insights into evolution.

1.2 Sequence Analysis

Sequence analysis is the processing of biological sequences by means of bioinformatics algorithms and data structures. The typical objective of sequence analysis is to answer questions from biological or medical research. In this section, we will present several examples of sequence analysis tasks as well as some common tools for sequence analysis.

1.2.1   Sequence Assembly

All known methods for determining biological sequences are only capable to directly decrypt sequences of limited lengths. We need sequence analysis algorithms to produce longer sequences. For example, the first sequencing of the human genome (Venter et al. 2001; International Human Genome Sequencing Consortium 2001) was based on the chain-termination sequencing technique (Sanger et al. 1977) which is only capable to produce sequence reads shorter than a thousand nucleotides. For longer sequences a procedure called *shotgun sequencing* proved to be viable, and it is able to determine even the sequences of whole eucaryotic genomes (*whole-genome shotgun sequencing,* Staden 1997; see also Istrail et al. 2004). This method randomly samples and sequences fragments from the DNA such that on average any part is covered several times. The resulting sequence *reads* are then *assembled* by means of sequence analysis methods to get the complete sequence. This is not trivial because errors occur during the sequencing of the reads, it is not clear which strand of the double-helix the read stems from, and because genomes are highly repetitive.

A first step in sequence assembly is usually to compute *overlap alignments* between the reads, for example by a *dynamic programming* alignment algorithm (see Section 8.5.4 on page 124). If two reads significantly overlap, then they putatively originate from the same location. For large numbers of reads, the computation of all needed overlap alignments could be accelerated by applying filtering. For example one could limit the search for overlapping candidates to those pairs of reads that share at least a given number of common *q*-grams (see Section 11.2.1 on page 187). The question is then how to derive from the overlap information the complete sequence. Usually several processing steps are necessary for computing a final consensus sequence (see, e.g., Huson et al. 2001). Sequence assembly is especially hard for DNA that contains long repeats, since all reads that stem from repetitive regions cannot be definitely assigned to a single position. In this situation it may be helpful to apply *double-barrel* shotgun sequencing, that is to sequence both ends of fragments that have a fixed length of several thousand nucleotides; see Figure 2. From that we get pairs

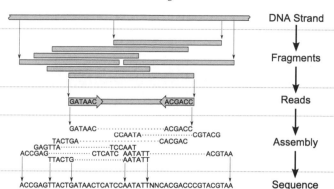

Figure 2: **Double-Barrel Shotgun Sequencing.** Fragments of a DNA strand are sequenced from both sides. The sequence assembly problem is to reconstruct the sequence of the DNA from the sequences of the reads.

of sequence reads with a certain distance in between, so if one of these reads falls into a repeat region, it may still be possible to determine its real position relative to its mate read. There are several de-novo sequence assemblers available; some of the more recent tools can even handle the rather short reads that are generated by next-generation sequencing technologies (e.g., Dohm et al. 2007; Zerbino and Birney 2008).

1.2.2 BLAST: Finding Similar Regions

Searching is probably the most basic operation in sequence analysis, and a program called Basic Local Alignment Search Tool (BLAST, Altschul et al. 1990) has become the most widely used sequence analysis tool in bioscience. Comparable tools like for example FASTA (Pearson 1990) or BLAT (Kent 2002) are less popular. BLAST is a heuristic for finding optimal *local alignments* (Section 10.1) in two input sequences a and b. That means it searches similar substrings in a and b, where the *similarity* (Section 8.3.2 on page 114) between two strings is defined by the score of an optimal *alignment* between them. The longer the strings are and the less they differ, the higher is this score. BLAST does not only compute similar regions and scores, but it also estimates a statistical significance, i.e., it computes the probability for finding similar regions of a certain length in two *uncorrelated* sequences a and b simply by chance (Karlin and Altschul 1990). If this probability is very small, then we can conclude that the regions in a and b are probably *correlated* and there must be a reason for their similarity. Since BLAST runs very fast, it has turned out to be an extremely useful tool in practice, and therefore the paper by Altschul et al. (1990) became one of the most cited publications in science history. There are several variants (e.g., *blastp* for Proteins or *blastn* for DNA) and implementations (e.g., NCBI-BLAST or WU-BLAST) of the tool.

In short, the algorithm works as follows: BLAST first searches for *seeds* (Section 8.6.1), which are highly similar regions as for example exactly matching substrings of a certain length q (q-grams). This means that BLAST finds only those local alignments that contain a seed, so it will find more alignments if the seed length q is reduced, although this will also slow down the search. The seeds may be found, e.g., by an automaton (Section 12.1) or a q-gram hash index (Section 11.2). Each seed is extended in both directions by a *X-drop extension* (Section 10.2.1). The resulting local alignment is stored if it exceeds a certain level of quality. In the end, the best local alignments are printed out.

SeqAn supports data structures and algorithms both for finding and extending seeds (Sections 8.6.1 and 10.2), as well as functions for parsing the output of standard BLAST tool.

1.2.3   CLUSTAL W: **Aligning Multiple Sequences**

Among the most important tasks in sequence analysis is the *alignment* of sequences (Section 8.2): The sequences written one below the other form the rows of a matrix, and blank characters are inserted into these rows such that similar parts of the sequences are grouped together. For similar sequences an alignment algorithm usually groups matching characters in one column using only a small number of gaps. For such alignments the *score* of the alignment (Section 8.3.1) is usually high. Alignments may explain a lot about sequences, since they reveal both the similarities between them and the small differences within these similarities. If the sequences for example originate from different species, then the optimal alignment can be used to infer their phylogenetic relationship.

We will show in Section 8.5.1 how to compute an optimal alignment between two sequences by *dynamic programming* in quadratic time. Unfortunately, the runtime grows exponentially with increasing numbers of sequences, and it was shown that the alignment problem is NP-hard (Wang and Jiang 1994), so practical tools for aligning multiple sequences are based on heuristics. One of the most common tools of this kind is CLUSTAL W, which applies a *progressive* approach; see Section 8.5.5. The tool works in three steps (see also Algorithm 5 on page 126):

(1) The pairwise distances between the sequences are computed, either by counting common q-grams or by aligning them. The result is stored in a distance matrix.

(2) From this distance matrix, a *hierarchical clustering* algorithm like UPGMA (e.g., Sneath and Sokal 1973) or *neighbor-joining* (Saitou and Nei 1987) computes a rooted binary *guide tree*. The leaves of the guide tree correspond to the sequences that are to be aligned.

(3) The sequences are aligned following the guide tree from the leaves to the root. At each inner vertex of the tree, the multiple alignment between all leaves below this vertex is computed by *aligning the alignments* of the two child vertices, see Figure 3.

This method is *greedy*, because once two sequences are aligned, then this alignment will be retained until the algorithm stops. A gap that is inserted will never be moved or removed again, and new gaps always affect the whole column of the alignment, thus any error that occurs in the early stages of the algorithm will be propagated to the end of the computation. As a remedy, CLUSTAL W applies a clustering algorithm to construct the guide tree, because this way similar sequences are joined earlier than distant sequences, and similar sequences are more likely to be aligned correctly.

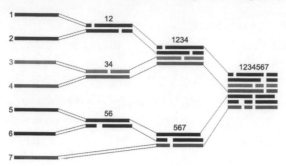

Figure 3: **Progressive Multiple Sequence Alignment.** The sequences on the left are aligned following a guide tree. Each vertex aligns the alignments from its child vertices, so the alignment on root vertex contains all input sequences.

SeqAn also offers progressive alignment algorithms that follow the improved T-Coffee tool (Notredame et al. 2000); see Section 8.5.5.

1.3   Software Libraries

One main goal of bioinformatics is to devise algorithms and develop software tools for biological and medical research. In this section, we will discuss how software libraries may improve the development of tools for sequence analysis. A *software library* is a set of reusable components, i.e., data structures and algorithms that use and manipulate these data structures. A component is *reusable*, if it can be used in different programs and by different programmers.

This is illustrated in Figure 4 that shows the core components of four tools for genome alignment: LAGAN (Brudno et al. 2003) MUMmer (Kurtz et al. 2004) MGA (Hohl et al. 2002) Mauve (Darling et al. 2004). All these tools perform the following three steps: (1) search for seed fragments, (2) compute an optimal chain from these seeds (Section 8.6), and finally (3) close the gaps between the seeds. Obviously these tools apply similar building blocks, like (enhanced) *suffix arrays* (Section 11.3) for seed find-

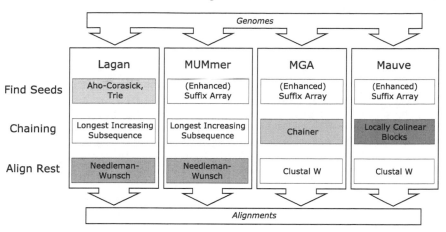

Figure 4: **Components of Genome Alignment Tools.**

ing, *longest increasing subsequence* (Section 12.2.2) for chaining, or the Needleman-Wunsch algorithm (Section 8.5.1) for aligning the spaces between the chained seeds. In the example it is evident that the developers of these tools would have profited from having a software library at hand that provides efficient implementations of the named algorithmic components.

1.3.1 Benefits from Software Libraries

Software development may benefit from software libraries in several ways: Their application *simplifies* the implementation of software, if a programmer can employ ready-to-use components from the library instead of re-implementing every part of the program, or to bother about the implementation details of the actually used algorithms or data structures. This *accelerates* the development process, which allows the program to be earlier on the market or – in the case of a bioinformatics tool – in the laboratories. These time savings *reduce the costs* of software development. Software libraries may also improve the *quality* and *robustness* of the resulting code, because the components of a library are widely used and therefore usually well tested. Moreover, the application of library components may also improve the program's *performance*, since libraries are intended to be used many times, so they may offer advanced but fast algorithms or data structures the implementation of which would not pay out for a single project. This also shows that software libraries support *algorithm design* by providing benchmarks for well known problems and acting as test environments in which a new algorithm may prove its correctness and possibly its superiority to previous approaches. A well designed library invites the user to *play around* with algorithms and data structures and thus promotes a fast testing of new *algorithmic ideas*. Finally, the best way for a new algorithm to arrive in software development is to publish it in a widely used software library, so libraries may help to close the gap between theory and practice in algorithmic research.

1.3.2   Software Library Examples

We will now briefly present two software libraries that are good examples for the usefulness of libraries in software development and that influenced the design of our library SeqAn.

1.3.2.1   LEDA

The Library of Efficient Data Types and Algorithms (Mehlhorn and Näher 1999) is a C++ library for combinatorial and geometrical computing. Since it was proposed in 1989 by Mehlhorn and Näher, LEDA grew to an extremely comprehensive software library. The range of its functions contains basic containers like arrays, lists, or sets, and it provides number types and graphs as well as algorithms and data structures for linear algebra, geometry, compression, and cryptography. LEDA was designed from the beginning to advance the transfer of theoretical algorithmic knowledge to practical tool programming, and in this respect it became a model for software libraries like SeqAn.

1.3.2.2   STL

The Standard Template Library (Stepanov and Lee 1995) is a C++ template library of basic containers and algorithms. With some changes and extensions, it became a part of the C++ standard library (ISO/IEC 1998; Josuttis 1999), so we will call this part of the C++ standard library the STL. The STL was one of the first C++ libraries that applied *generic programming* (e.g., Austern 1998; see Section 4.2), and it demonstrates that this programming paradigm is capable of implementing flexible and performant libraries. The algorithms of the STL access and modify the contents of the containers by *iterator* objects; further key concepts of the STL are *functionals*, i.e., objects that implement the parenthesis operator (), and *type traits* that resemble the *metafunctions* that we use in SeqAn (Section 4.5). The *library design* of SeqAn (Chapter I) can be regarded as an advancement of the techniques that were introduced into library design by the STL.

Chapter 2

SeqAn

Here we summarize the goals and design of the software library
SeqAn, our generic C++ template library of efficient data types and
algorithms for sequence analysis (see Figure 5). The development
of SeqAn has pursued two main goals, namely:

(1) Enabling the rapid development of efficient *tools* for sequence
analysis.

(2) Promoting the design, comparison, and testing of *algorithms*
for sequence analysis.

SeqAn accelerates the development process of tools and algorithms, and improves the quality and performance of sequence
analysis software. In addition, it provides an experimental platform for algorithm engineering and closes the gap between state-
of-the-art algorithmic techniques and the actual algorithmic components used in software tools (Section 1.3). SeqAn is the first
software library with this ambition that was actually realized.

2.1   Design of SeqAn

SeqAn was designed to promote (1) high *performance* of the provided components, (2) *simplicity* and usability of the library's handling, (3) *generality* of data types and algorithms such that they
are widely applicable, (4) the definition of special *refinements* of
generic classes or algorithms, (5) the *extensibility* of the library,
and (6) easy *integration* with other libraries (Chapter 3).
We decided to implement the library in C++, since C++ provides
language constructs that allow to achieve our design goals (see

Figure 5: **The SeqAn Logo.**

Section 4.1). The unique library design of SeqAn is based on (1) the *generic programming* paradigm, (2) a new technique for defining type hierarchies called *template subclassing*, (3) *global interfaces*, and (4) *metafunctions*, which provide constants and dependent types at compile time (Chapter 4). Our design differs from common programming practice, in particular SeqAn does *not* use object-oriented programming (Section 4.3.2). However, the library benefits greatly from our approach, and all design goals are met.

2.2   Contents of SeqAn

SeqAn is a comprehensive library that was intended to cover a wide range of topics of sequence analysis. It offers a variety of practical state-of-the-art algorithmic components that provide a sound basis for the development of sequence analysis software. This includes: (1) data types for storing strings, segments of strings and string sets, as well as functions for all common string manipulation tasks including file input/output, (2) data types for storing gapped sequences and alignments, and also algorithms for computing optimal sequence alignments, (3) algorithms for exact and approximate pattern matching and for searching several patterns at once, (4) algorithms for finding common matches and motifs in sequences, (5) string index data structures, and (6) graph types for many purposes like automata and alignment graphs, as well as many algorithms that work on graphs (Part II).

SeqAn offers several alternative implementations for all core data types like strings, string sets, alignments, graphs, and indices. It also provides a variety of different algorithms for central tasks like pattern matching, motif finding, or the alignment of sequences. The user can therefore select the variant that fits best to the actual application.

SeqAn was already applied for the development of several state-of-the-art software, e.g., Schulz et al. (2008), Weese and Schulz (2008) Rausch et al. (2008), Langmead et al. (2009), and Rausch et al. (2009). This demonstrates its usability. We will also substantiate this claim in Part III where we propose several applications of the library.

2.3 Testing

Each part of SeqAn was exhaustively tested. We used two testing strategies (see, e.g., Myers et al. 2004):

(1) **Unit Testing**: For each module of SeqAn exists a program which tests all data structures and functions that reside in this module. These are mainly *black-box* tests, i.e., the program does not inspect the actual implementation but only check the correctness of output generated by the tested library part. In many cases the input data is predefined within the test program; some other tests generate repeatedly random inputs and compare the outputs of alternative implementation.

(2) **Function Coverage Testing**: The extensive use of C++ templates in SeqAn raises a special testing problem: Usually C++ compilers perform only shallow syntax checks during the parsing of template code, so it is possible that a test

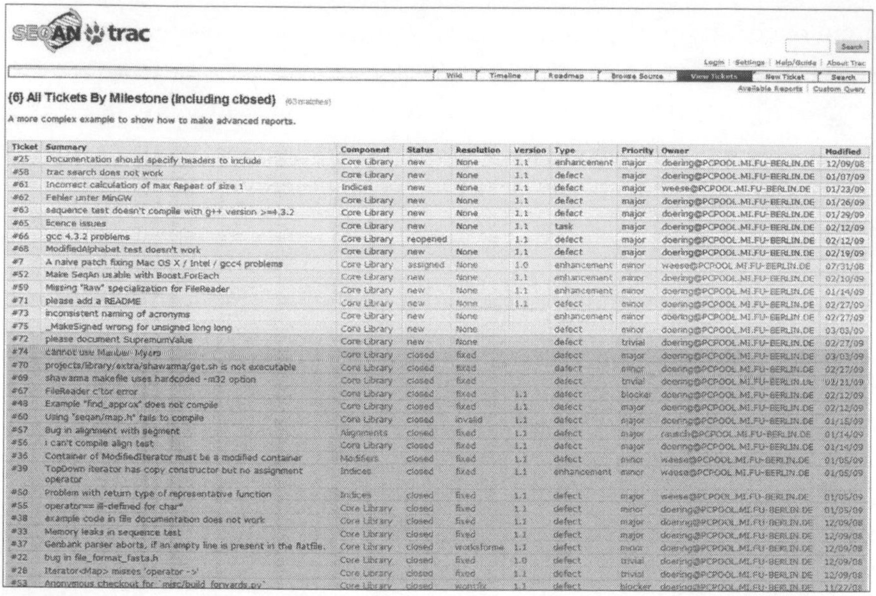

Figure 6: **SeqAn Trac.** The screen shot shows a list of messages and issues posted by SeqAn users from around the world.

program compiles correctly even if some templates contain syntax errors, just because these templates are never instantiated. We therefore apply a *white-box* testing method that ensures each template function to be instantiated at least once in the test. This is done by inserting the preprocessor macro SEQAN_CHECKPOINT at the beginning of each template function of the library, and maybe also in some further parts of the program for which we want to check that they are reached by the test. If testing is activated, this macro expands to a short piece of code that protocols the current source file and line of code. At the end of the test, the source files are scanned for SEQAN_CHECKPOINT and all occurrences that were never reached are reported. If testing is not activated, then the macro is defined to be empty, so SEQAN_CHECKPOINT has in this case no impact on the program's efficiency.

Since even the best testing cannot guarantee the correctness of a program, we used the open source error tracking system *Trac*[1], so the library's users can report their bugs and give suggestions for improvement; see Figure 6.

2.4 Documentation

The common documentation systems for C++ like *Doxygen*[2] are designed with regard to object-oriented programming, so we developed our own documentation system *DotDotDoc* (DDDOC), which is especially suited for documenting generic programming software. The documentation is deposited in C++ comments that are ex-

Figure 7: **SeqAn Documentation Using DDDOC.** The screen shot shows a part of the documentation for the class **String**.

---

[1]See `trac.edgewall.org`
[2]See `www.doxygen.org`

tracted from the library's source files using a Python (Lutz 2006) script. The format orientates on the XML documentation format[3] that is used for Microsoft C#, but it uses a simple human readable notation style instead of XML. DDDOC creates a heavily cross-linked and searchable documentation (see Figure 7) that extensively describes all public classes, specializations, functions and metafunctions available in SeqAn on HTML pages, which can be viewed in common HTML browsers. The SeqAn documentation also contains several tutorials and example programs. It can be downloaded from the SeqAn Web site and viewed online on www.seqan.de/dddoc.

2.5   Distribution

During the development process we took care to keep SeqAn compatible with multiple platforms. For that reason, we implemented a simple but powerful built-system that allows the compilation of applications and test programs using different compilers and operating systems. Our library now works on all common platforms, namely Microsoft Windows, Mac OS X, Solaris, and several Linux clones, and it was tested for Microsoft Visual C++ compilers (version 7 or above) and GCC compilers (version 3 or above).

SeqAn is an open source and free software published under the Gnu Lesser General Public License (*LGPL*) version 3.[4] This license allows the free use and distribution of the library also for commercial use. Both the library sources and the documentation can be viewed and downloaded from the SeqAn Web site www.seqan.de, which was designed to be the central place for all news and information about the project; see Figure 8. Besides detailed descriptions of SeqAn and its associated projects, this Web page also contains a bug tracker system that can be used to return feedback to the library's developers; see also Section 2.3.

---

[3]see msdn.microsoft.com
[4]See www.gnu.org/licenses/lgpl.txt

Figure 8: **SeqAn Web site.**

Chapter 3

Library Design

3.1 Design Overview

In this part, we discuss the *core library design* of SeqAn. We call it *core* design, because it answers very basic questions like: What are the strategies for organizing the functionality in the library? What is the general form of classes and functions? What language features are applied, and how are they used? The core design does *not* specify what classes and functions should be implemented in the library. This *detailed* design will be the topic of Part II, in which we will give a complete overview of the contents of SeqAn. Although the core design is not directly connected with the actual contents of the library, it is influenced by the kind of functionality the library offers. For example we observe that sequence analysis relies on rather simple but *generic* data structures like sequences (Chapter 7), alignments (Chapter 8), string indices (Chapter 11), and graphs (Chapter 12) which makes it amenable to the generic programming paradigm, whereas libraries consisting of less generic but very complex data structures would probably be better implemented in a more *object oriented* way.

The decision for an appropriate core design also depends on the intended *application* of the library. As we stated in Chapter 2, SeqAn has the purpose to facilitate the development of new sequence analysis tools, and it is an algorithm engineering platform for comparing and developing efficient data structures and algorithms. Both applications require that the components of the library run as fast as possible, so *performance* is one of the most important objectives during the library design phase. Considerations like this lead us in Section 3.2 to six *main goals* for the core design of SeqAn. In the following Chapter 4, we will discuss by what programming techniques these goals can be achieved. It turns out that only a few powerful techniques suffice. The mechanics of the resulting core design is then demonstrated by examples in Chapter 5.

3.2   Design Goals

3.2.1   Performance

A first – and maybe most important – objective for SeqAn is *performance*:

> *The library is designed to produce code*
> *that runs as fast as possible.*

Since data structures usually must fit completely into main memory to be fast, we also aspire to offer data structures with minimal space consumption.

While performance is of course a desired feature of any software, it plays a critical role in the competition between software tools. For example, some applications in bioinformatics involve huge problem instances which may take running times of several hours or even days, so a tool's speed can make the difference between a feasible and an infeasible experiment if computing resources are limited.

In sequence analysis the amount of data to be analyzed usually forbids the application of brute force algorithms even for very basic tasks like searching a pattern in a string or aligning two sequences. Hence, one has to resort to efficient data structures and algorithms that achieve the required speedup. A library can supply very complex algorithmic components, which are hard or costly to implement for tool designers. Nevertheless, a tool designer has always the option to solve the problem at hand by its own specialized code or to resort to ad-hoc solutions, instead of using the components in a library. Programming specifically for a given problem may even yield better results than using standard components, depending on the effort involved, hence it is crucial for the library components to be competitive in speed to specialized code.

Optimal performance is also crucial for algorithm engineering: No algorithm designer would be happy to sacrifice hard-earned speedups obtained from a clever algorithmic advance due to sub-optimal implementations. A comparison between competing algorithms would always be influenced by varying implementation qualities, so the best way to make it fair is to compare implementations that are as effective as possible and are based on the same algorithmic components.

The need for performance is also our main reason to choose C++ as programming language (Stroustrup 2000; ISO/IEC 1998), since carefully designed C++ code has best chances to outperform most alternative programming language (see Section 4.1).

Achieving a good performance affects the library in many respects. If there is a trade-off between speed or coding convenience, then performance is usually favored. For example, we omit time consuming parameter checks in the release build of the library.

3.2.2 Simplicity

The second main goal for the library design of SeqAn is *simplicity*. Software libraries should facilitate the development of software, and hence they need a clear organization of their functionality. Plain interfaces improve a library's usability and accessibility, make it easier for a potential user to evaluate the usefulness of the library, and reduce the training needed to use it. In addition the internal mechanisms of a library should never get too complex, since this would slow down the development process of the library, complicate its maintenance, and it could be a source of hidden errors. A user of the library will become a victim of exotic compiler behavior, unreadable error messages, or inconsistencies in the language standard to the same extent the library uses elaborated language features.

Our goal is therefore:

All parts of the library are constructed and applicable as simply as possible.

We feel confident that the application of SeqAn is in fact simple, although this is always in the eye of the beholder, and in this book we will demonstrate the ease of use in a multitude of short code examples.

3.2.3 Generality

The next goal of SeqAn is *generality*: Library designers cannot completely anticipate all applications a library will actually be used for, so it is advisable to keep it as general as possible. A library that is useful in many circumstances has better chances to be used. Also, the more probable it is that a library can be re-used in future occasions, the more it pays for a user to get accustomed to it. Hence our goal is:

> *All parts of the library are applicable*
> *in as many circumstances as possible.*

General components are more intuitive to describe and easier to understand than data structures and algorithms that are only usable for a few individual cases, so generality also supports the *simplicity* (Section 3.2.2) of the library. A good starting point for finding general components is to identify common elements in different tools, as described in Section 1.3.

Generality also means that we try to avoid redundancy in the library: If, for example, one algorithm can work on different types of classes, then it should not be re-implemented for each class, but only once for all classes, in a single piece of code. This makes the library more compact and easier to maintain.

We will explain in Section 4.2 how *generic programming* enables us to create data structures and algorithms that work on a variety of types, for example how to implement strings of arbitrary alphabets, or algorithms that work an any kind of string.

3.2.4 Ease of Refinement

A good strategy for augmenting performance (Section 3.2.1) is to implement *specializations*: Sometimes the implementation of a function can be significantly improved, if we rely on a special context or the presence of some constraints. For example, searching an array is much faster if the values are sorted, thus it is advisable to define both a general but relatively slow linear search algorithm for unsorted arrays, and additionally a fast binary search algorithm for sorted arrays. A *specialization* overloads the *general* solutions (Section 3.2.3) for a special case, and the specialization can also be overloaded for an even more special case, so in the end we get a hierarchy of refinements.

The ideal library concept therefore fulfills the following rule:

> *Whenever a specialization is reasonable,*
> *it is possible to integrate it easily into the library.*

To integrate means that the new specialization works seamlessly together with the rest of the library, and that it can be applied in the same way as already existing alternatives. Our design therefore supports *polymorphism*, i.e., that the same interface may be realized by several implementations. This also promotes the simplicity (Section 3.2.2) of the library.

We will see in Section 4.3, how *template subclassing* enables us to implement specializations in a way that the C++ compiler always uses the most appropriate – i.e., the most special variant.

3.2.5 Extensibility

A classical slogan of good programming is the so-called *open-closed principle*, which states that a program should be open for extension but closed for modifications. We call this feature *extensibility*:

> *The library can always be extended*
> *without changing already existing code.*

Extending the library means to overwrite default behavior by defining new specializations (Section 3.2.4), or to add completely new functionality to the library. Extensibility is important both during the implementation of the library, because it simplifies its construction, and also for a user who wants to adapt the library to his needs.

3.2.6 Integration

It is often reasonable to use several libraries at once. This means that the libraries must be able to collaborate with each other:

> *The library is able to work together with*
> *other libraries and built-in types.*

This includes that SeqAn obeys some *rules of coexistence,* for example, to use its own namespaces `seqan` in order not to contaminate the global namespace, or not to define preprocessor macros that could conflict with code of other libraries. Moreover, we aim at providing means for a direct integration of external libraries: For example, string classes are provided not only by SeqAn (see Section 7.3) but also by many other libraries like the STL (Plauger, Lee, Musser, and Stepanov 2000) or LEDA (Mehlhorn and Näher 1999), and strings can also be stored in `char` arrays, so-called *C-style strings.* It would be of great advantage, if we could implement algorithms that work on all these kinds of string.

We will explain in Section 4.4 how the SeqAn library design supports this kind of integration by using small global functions or metafunctions – so-called shims – to adapt external interfaces to the needs of SeqAn.

Chapter 4

Programming Techniques

In this chapter, we discuss the main techniques used in SeqAn to achieve the design goals that we described in Section 3.2, namely *generic programming* (Section 4.2), *template subclassing* (Section 4.3), *global interfaces* (Section 4.4), and *metafunctions* (Section 4.5). The combination of these four techniques forms the *core design* of SeqAn. In Section 4.6 we will propose further programming techniques that we apply in SeqAn. We start with discussing the reasons for using the programming language C++.

4.1 The C++ Programming Language

The programming language C++ was proposed by Bjarne Stroustrup in 1983 (see Stroustrup 2000) as an extension of the procedural and imperative programming language C (Kernighan and Ritchie 1988). SeqAn relies on ISO/IEC standard conform C++ (ISO/IEC 1998) that is supported by several compilers like the GNU C++ compiler (Griffith 2002) or the Visual C++ compiler (Visual C++ 2002).

The programming language C was designed as "a relatively *low level* language" so that "the data types and control structures provided by C are supported directly by most computers.[1]" Although C was designed to be independent from a particular architecture, it is in effect rather machine-oriented, which means that C programs match the capabilities of present computer architectures and have therefore best chances to run fast. During the compi-

---

[1]See (Kernighan and Ritchie 1988), pages 5–6.

lation of C source code, the compiler may also apply optimizations to achieve further speed-ups. C++ enhances C by concepts like *object-oriented programming* and *generic programming* (Section 4.2), and though some of these new features (e.g., *virtual functions*) entail pitfalls to slow down the resulting programs, carefully employed C++ achieves in general the same performance as C. Due to the prevalence of C/C++ in the last decades, the co-evolution of computers and compilers made these languages probably the best choice for high-performance applications. We decided to implement SeqAn in C++, because *performance* is among our main goals (Section 3.2.1) and the extended features of C++, namely *templates* (ISO/IEC 1998, section 14), are well suited to attain an excellent *library design*.

There are prominent examples of C++ software libraries in the area of algorithm engineering like LEDA (Mehlhorn and Näher 1999) and CGAL (Fabri et al. 2000), many common software tools for sequence analysis like NCBI Blast (Altschul et al. 1990) are implemented in C++.

4.2   Generic Programming

SeqAn adopts *generic programming*, a paradigm that was proven to be an efficient design strategy in the C++ standard (ISO/IEC 1998). The standard template library (STL) (Plauger et al. 2000) as part of the C++ standard is a prototypical example for generic programming. Generic programming designs algorithms and data structures in a way that they work on all types that meet a minimal set of requirements. An example for a generic data structure in the STL is the class `vector`: It is a container for storing objects of a type `T` that are *assignable* (ISO/IEC 1998, section 23.1), which means that we can assign one instance `s` of `T` to another instance `t` of `T`, i.e., the code `T t = s` is valid. This kind of requirement to the interface of a type `T` is called a *concept*, and we say that a type `T` *implements* a concept, if it fulfills all requirements stated by that concept; for example the concept *assignable* is implemented by all built-in types and every class that has both a copy assignment operator and a copy constructor. Generic programming has two implications: (1) Data structures and algorithms work on *all* types T that implement the relevant concept, i.e., relevant is not the type T *itself* but its interface, and (2) this concept is *minimal* in the sense that it contains only those requirements that are essential for the data structure or algorithm to work on T. This way data structures and algorithms can be applied to as many types as possible, and hence generic programming promotes the *generality* of the library (see Section 3.2.3).

Generic data types and algorithms can be implemented in C++ using *templates* (ISO/IEC 1998, section 14). A class template parameterizes a class with a list of types or constants. For example, a declaration for the class `vector` could be:

```
template <typename T> class vector;
```

where T stands for the *value type*, i.e., the type of the values that will be stored in `vector`. The template is generic, it can be applied to any type T. For example, a vector for storing `int` values is instantiated by:

```
vector<int> my_vector;
```

That is we use `int` as template argument for T, and the result of the instantiation is an object `my_vector` of the *complete type* `vector<int>`. The compiler employs the same template, i.e., the same piece of code, for different template argument types. The compilation succeeds if the applied template argument type supports all uses of the parameter T within the template code, so the C++ template instantiation process implies the *minimality* of the concepts.

Listing 1 shows an example for a generic algorithm. The function template `max` can be applied for two objects `a` and `b` of any type T that is *assignable* and can be compared using the < operator. The compiler may implicitly derive the type T from the given function arguments, for example `max(2, 7)` calls the instantiation of `max` for T = `int`.

```
template <typename T>
T max(T a, T b)
{
   if (a < b) return b;
   else return a;
}
```

Listing 1: **Example of a Generic Algorithm**. The function template `max`
   returns the maximum of two values a and b, where a and b could be from
   any suitable type T.

4.3   Template Subclassing

A generic algorithm that is applicable to a type T needs not to
be optimal for that type. The algorithm `find` in the standard
library (ISO/IEC 1998, section 25.3.1.1) for example performs a
sequential linear time search and is therefore capable of finding
a given value in any standard compliant container. However, the
container `map` was designed to support a faster logarithmic time
search, so the algorithm `find` – though applicable – is not optimal
for searching in `map`. This shows that sometimes a special algorithm could be faster than a generic algorithm. Hence, in order to
achieve better performance (Section 3.2.1), we require our library
(see Section 3.2.4) to support *refinements* of algorithms. A special
version is only useful if it really allows a speedup in some cases,
and only in this case it will actually be implemented. Therefore
we assume that for a given case always the *most special* applicable
variant is the best, where we have to assure that there is always a
definite *most special* candidate according to the C++ function overload resolution rules (ISO/IEC 1998, sections 13.3 and 14.5.8).

Since one of our goals is *simplicity* (Section 3.2.2), and since it
could be rather demanding for the user to find out the best algorithm out of various alternatives, we decide to apply *polymorphism*, that is, all alternative implementations of an algorithm
support the same interface. So we can write `find(obj)` for any

```
template <typename TValue, typename TSpec> class Container
{
    // generic container
};

struct Map;

template <typename TValue> class Container<TValue, Map>
{
    // special map container
};
```

```
template <typename T> void find(T &)
{
    // most general: works for all types
}

template <typename TValue, typename TSp>
        void find(Container<TValue, TSp> &)
{
    // more special: works for all containers
}

template <typename TValue> void find(Container<TValue, Map> &)
{
    // even more special:works   only for maps
}
```

Listing 2: **Template Subclassing Example**. Note that SeqAn does not
implement a class Container.

container type `obj`, and this invokes the most suitable implementation of `find` depending on the type of `obj`. Listing 2 gives an example of this idea: The map is implemented in the specialization `Container<Map>` of the generic class `Container`. Since the subclass is specified by choosing a template argument, we call this approach *template subclassing*.

In the lower part of Listing 2, we see different levels of specificity for `find` algorithms: The first is applicable for any type – therefore we call it the *most general function* – the second for instances of `Container`, and the third only for `Container<Map>` objects. The rules of C++ *function overload resolution* (ISO/IEC 1998, section 13.3) assures that the correct variant is called.

We need not to end the specialization on the level of `Container<Map>`. Suppose that we define `Map` as a class template with a template parameter `TSpec`, then we can also implement special variants of maps, for example `Container<Map<Hashing> >`. This way, we can define specialization hierarchies of unlimited ramification.

Note that we make no demands on template argument types like `Map` that are used for `TSpec`, in fact any type that is merely *declared* can be used as template argument, so its only important aspect is to act as a switch between different specializations of `Container`. We call a class that is intended merely to serve as switch a *tag class*. Tag classes can also be used to switch between different modes of a function (see Section 4.6.2).

4.3.1   Template Subclassing Technique

The technique of *template subclassing* may be summarized as follows:

- The data types are realized as default implementation or specialization of class templates, e.g., `Class`, which have at least one template parameter `TSpec`.

- Refinements of `Class` are specified by using in `TSpec` a *tag class*, e.g., `Subclass`, that means they are implemented as class template specializations `Class<Subclass>`.

- Whenever further refinements may be possible, we declare the tag classes as class templates with at least one template parameter `TSpec`, in which more tag classes can be used. For example we may implement a class template specialization `Class<Subgroup<Subsubgroup<...> > >`. This way, we can reach arbitrary levels of specialization.

- Algorithms can be implemented for each level of specialization. If multiple implementations for different levels of specialization exist, then the C++ function overload resolution selects the most special from all applicable variants.

Note that we only need to define a class template specialization `Class<Subclass>` explicitly, if the *members* of the new refinement differ from the members of its *parent class* `Class`. We will see in Sections 4.4 and 4.5 that member functions and member types play a minor role in SeqAn, so in many cases an actual specialization of the class template is not needed.

4.3.2   Comparison to Object-Oriented Programming

Template subclassing resembles class derivation in standard object-oriented programming. In the spirit of the object-oriented terminology we can say that the class `Container<Map>` was *derived* from the general class `Container`, since all algorithms that are defined for `Container` also work for `Container<Map>` and are therefore *inherited*. Let us compare this to object-oriented programming: The method `find` is inherited from the base class `Container` to all derived classes, but it was overloaded for the class `Map` which defines its own `find` method.

This approach has two drawbacks: The first disadvantage is that the method `find` is not a generic algorithm according to the definition in Section 4.2 but a member function. Its usage has the form `obj.find()`, which differs completely from the application of the generic algorithm that would be called by `find(obj)`. If we try to achieve the *refinement* goal (Section 3.2.4) this way, and if we therefore define both global functions and member functions,

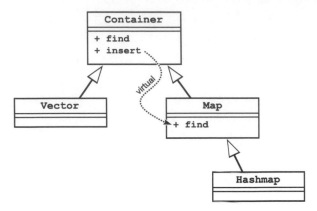

Figure 9: **Object-Oriented Example.**

then the handling of the library gets more complicated, so we lose *simplicity*.

Object-oriented programming has another drawback, see Figure 9. Suppose that we want to define a function `insert` that adds a new value to a container, and that `insert` calls `find` in order to check whether the container already holds the given value. Since `insert` can be applied for all containers, we implement it as a member function of `Container`, so it is inherited also by `Map`. If we call `insert` to insert a value into a `Map` object, then `insert` should use the correct `find` function, i.e., the special *logarithmic search* that was defined in `Map` rather than the general *linear search* defined in `Container`. Therefore `find` has to be declared *virtual* (ISO/IEC 1998, section 10.3), which means that each call of `find` costs an additional overhead. Virtual function calls are indirect via the lookup in a table of function pointers, and they are therefore more complicated than *ordinary* function calls; in contrast non-virtual functions have the advantage that C++ compilers may use function inlining to completely save the overhead for calling them. The application of virtual functions in this case therefore reduces the *performance*.

We conclude that template subclassing better fits our design goals than object-oriented programming. With template subclassing we can implement *polymorphic generic algorithms*, which means that the refinements of algorithms have the same interface as their generic counterparts. Template subclassing also needs no virtual functions but relies completely on *static function binding*, that is the compiler determines during compile time which function is called and can therefore apply optimizations to improve performance.

4.4 Global Function Interfaces

A *global function* in C++ is a function that was declared in namespace scope (ISO/IEC 1998, section 3.3.5). In contrast to that, *member functions* are defined in the scope of a class. *Object-oriented programming* prefers member functions, because they work as *methods* which can be inherited and overloaded during the class derivation. *Generic programming* on the other hand also applies *global* function templates for implementing *generic* algorithms (see Section 4.2). In Section 4.3 we saw how we can use global function templates to implement algorithms for different levels of specialization. SeqAn relies on global functions anyway, and therefore the following design decision seems rather natural: SeqAn abstains from accessing objects via member functions as far as possible, that is, all functionality in SeqAn is accessed via global functions, with the exception of functions that must be members due to language restrictions, like constructors, destructors, assignment operators, bracket and parenthesis operators, and conversion functions. Since global functions substitute those member functions that would otherwise form the *interface* of a class, we call all functions that accept instances of a class as arguments the *global function interface* of this class. For example, to determine the length of a string `str` in SeqAn, we call a global function `length(str)` instead of a member function `str.length()`. Using global interfaces is a main feature of SeqAn, and we will see in Section 4.5 that SeqAn also applies a global interface for accessing types.

Obviously the most direct way to achieve global function interfaces is simply to implement the functionality in global functions. This is also a prerequisite for using *template subclassing*, so most implementations in SeqAn actually reside in generic global functions, especially as far as this concerns the classes defined in the library. Alternatively, one can implement the functionality in a member function, and then call it via a *shim* (Wilson (2004), ch. 20), i.e., a small global function that acts as a *wrapper* for the member function. Shims are a good way to create new global interfaces for already existing data structures or for built-in types. For example, SeqAn contains several global functions `length` that work on the `basic_string` class from the C++ standard library or on zero terminated `char` arrays (so-called *C-style strings*). So for determining the length of string `str` we can always call `length(str)` regardless of whether `str` is an instance of the SeqAn class `String` (Section 7.1), or a `basic_string`, or an array of `char`. See Section 5.1 for more details about this example.

4.4.1   Advantages of Global Interfaces

Overall there are some good technical reasons to favor global functions over member functions (see also Czarnecki and Eisenecker 2000, 6.10.2):

- Global functions greatly support the *open-closed principle*, i.e., they favor the *extensibility* (Section 3.2.5): New functions can be added to the library at any time without changing the library's code. This holds true for new specializations of already existing functions, as well as for completely new functionality. Moreover, it is possible to encapsulate the declarations of global functions in different header files and include them only if they are needed.

- The shim technique allows us to adapt arbitrary types to uniform interfaces, so using global functions is a good way to attain *integration* (Section 3.2.6) of the library with other libraries and built-in types.

- The difference between global *algorithms* and non-global member functions, as they are used for example in the standard template library, can be somewhat confusing, especially if there exist algorithms and member functions with the same name. Therefore, adapting global functions *simplifies* the library (Section 3.2.2). Moreover, global functions do not assume that one special function parameter acts as the *owner* of the function, so they may sometimes be more intuitive, e.g., when modeling symmetric operations like a matrix multiplication, which has no preference to be a member of the first or the second matrix.

- As we saw in Section 4.4, the obvious way to implement generic algorithms in C++ is to use global function templates, so global functions help to achieve a maximal *generality* (Section 3.2.3) of the library.

4.4.2   Discussion

Many programmers will probably at first be skeptical about our preference of global functions, since our approach contradicts common rules of object-oriented programming, so we now discuss some possible objections.

Missing Protection

Global functions lack a protection model: They cannot be private or protected, and they cannot access private and protected members of a class.

We addressed this problem by establishing rules of good coding practice. The main reason for a protection model is to prevent the programmer from accessing functions or data members that are intended for internal use only. A simple substitution for this feature is to establish clean naming conventions: We state that a '_'-character within an identifier indicates that it is for internal use only. Global functions can access private members of a class C if they are declared to be *friends* of C, but our experience showed that this approach is too inconvenient in practice, so we instead decided to declare data members to be public, but only functions that belong to the core implementation of C are allowed to access them by convention.

Possible Ambiguities

One could argue that we risk ambiguities when we define functions for several classes with the same name. Suppose that a class `String` and another class `Tree` should both support a function `length`. Then we simply implement two functions `length(String & str)` and `length(Tree & str)`, and this will work since both functions have different argument lists. A problem may only arise if multiple functions could be applied for the given arguments, in this case, we have to take care that there is always a *best* alternative according to the C++ rules for function overload resolution (ISO/IEC 1998, sections 13.3 and 14.5.8).

Handling of Namespaces

To avoid conflicts with other libraries, SeqAn defines all data types and functions for public use in the special namespace `seqan`. Nevertheless we need not to specify this namespace whenever we call a global function, because C++ specifies a rule for *argument-dependent name lookup*, also known as *koenig lookup* (ISO/IEC 1998, section 3.4.2), which means that if the compiler looks for the actual function `length` that is called by `length(str)`, then it also searches the namespace in which the type of the argument `str` was defined, so if `str` is an instance of the class `String`, then the matching function `length` is found, since both the function and the class are defined in the same namespace `seqan`; see Listing 3. A function that gets arguments from different namespaces may cause ambiguities, so we decided not to use several namespaces in SeqAn.

Inheritance and Dynamic Binding

One may think that global functions are not inherited during class derivation, but in fact they are. Suppose that we have two classes `Base` and `Derived` and the second is derived from the first, then all functions that work for `Base` will also work for `Derived`. Nevertheless we do not make use of this observation in SeqAn, because we apply *template subclassing* instead; see Section 4.3.

Admittedly, global function cannot be virtual, but we will see that

```
namespace seqan
{
   class String { ... };
   size_t length(String &) { ... }
}

seqan::String str;
length(str);            //no namespace qualification needed
```

Listing 3: **Koenig Lookup Example**. We do not need to specify the namespace `seqan` when calling the function `length` from outside the namespace,
   because it is found by *argument-dependent name lookup*: The compiler
   searches in the namespace **seqan** for `length`, since the argument **str** was
   defined there.

template subclassing (Section 4.3) can substitute object-oriented
polymorphism in many cases, and since we use *static binding* instead of *dynamic binding*, our approach is much more efficient. If
dynamic binding is indispensable, one can still use virtual functions and call them via global functions.

Performance Overhead

In general, the overhead for calling global functions and (non-
virtual) member functions is the same. Shims are very small functions and will usually be inlined, so we do not need to expect
that shims affect the performance of a program if we apply an
optimizing compiler.

4.5   Metafunctions

Generic algorithms usually have to know certain types that correspond to their arguments: An algorithm on containers may
need to know which *type of values* are stored in the string, or
what kind of iterator we need to access it. The usual way in
the STL (Austern 1998) is to define the value type of a class like

vector as a *member typedef* of this class, so it can be retrieved
by vector::value_type. Unfortunately member typedef declarations have the same disadvantages as any members: Since they
are specified by the class definition, they cannot be changed or
added to the class without changing the code of the class, and it is
not possible in C++ to define members for built-in types. What we
need therefore is a mechanism that returns an output type (e.g.,
the value type) given an input type (e.g., the string) and that
thereby does not rely on members of the input type, but instead
uses some kind of global interface. Such task can be performed by
metafunctions, also known as type traits (Vandevoorde and Josuttis 2002, chapter 15). A *metafunction* is a construct to map some
types or constants to other entities like types, constants, functions,
or objects at compile time.

We use class templates to implement metafunctions in C++. Listing 4 shows an example for the definition and application of a
metafunction Value for determining the value type of containers.
The code defines Value for the Container class from Listing 2
and for C++ arrays. The returned type is defined as Type, so
Value<T>::Type is the value type of a container class T. The
generic algorithm swapvalues can be applied for both kinds of
data type for swapping the first two values stored in a container.

The metafunctions we propose here constitute a global interface
for accessing types, so they also share most of the advantages listed
in Section 4.4.1.

Metafunctions can also be used to define additional *dependent
types* that are not specified via template arguments. For example,
SeqAn offers the metafunction Size which specifies the appropriate type for specifying memory amounts (e.g., for storing lengths
of containers). This type is by default size_t, and it is hardly
ever changed by the user, so it is not worth to specify it in another template argument. Nevertheless it is possible to overwrite
the default with a new type, like a 64-bit integer (__int64) for
those container classes that provide extra large storage by defining a new specialization of the metafunction Size.

Our naming convention states that the return type of a metafunction is called Value. Another application of metafunctions is to

```
template <typename T> class Value;

template <typename TValue, typename TSpec>
class Value < Container<TValue, TSpec> >
{
    typedef TValue Type;
};

template <typename T, size_t I>
class Value < T[I] >
{
    typedef T Type;
};

template <typename T>
void swapvalues(T & container)
{
    typedef typename Value<T>::Type TValue;
    TValue help = container[0];
    container[0] = container[1];
    container[1] - help;
}
```

Listing 4: **Meta Functions Example**. The example class `Container` was
defined in Listing 2; it is not part of SeqAn.

define constants that depend on types. If a metafunction returns a constant, then this is called `VALUE`. For example, the metafunction `ValueSize` in SeqAn specifies for alphabet types the number of different values in the alphabet, so for `Dna` the metafunction call `ValueSize<Dna>::VALUE` returns 4.

4.6   Further Techniques

4.6.1   Metaprogramming

The name *metafunctions* (Section 4.5) stems from the fact that one can consider them as functions of a *metaprogramming language* that is interpreted by the compiler during the compilation process in order to produce the actual C++ code that is to be compiled afterwards. A metaprogram is processed during compile time and therefore does not burden the run time. One can do many things with metaprogramming (e.g., see Gurtovoy and Abrahams 2002), but since this technique is rather complicated and hard to maintain, we decided to use it only in limited circumstances. For example, SeqAn supports the metafunction `Log2` to calculate the integer logarithm of given constants; see Listing 5. This function is very helpful for example to compute the number of bits needed to store a value of a given alphabet type.

4.6.2   Tag Dispatching

Tag dispatching is a programming technique that uses the types of additional function arguments, called *tag arguments,* for controlling the *overload resolution,* which is the process of determining the function that is actually executed for a given function call (ISO/IEC 1998, sections 13.3 and 14.5.8). Since only the types but not the actual instances of the function arguments are relevant for overload resolution, a tag argument need not to have any members. Those classes are called *tag classes,* and we showed in Section 4.3.1 how tag classes are used in *template subclassing* to

```
template < int numerus >
struct Log2
{
    enum { VALUE = Log2<(numerus+1)/2 >::VALUE + 1 };
};


template <> struct Log2<1> { enum { VALUE = 0 }; };
template <> struct Log2<0> { enum { VALUE = 0 }; };
```

Listing 5: **Metaprogram Example.** This metaprogram computes the rounded up logarithm to base 2. Call Log2<c>::VALUE to compute $\lceil \log_2(c) \rceil$ for a constant value c.

select a data structure out of several alternatives.

Listing 6 shows how we can use tag classes to switch between different implementation alternatives of algorithms. The third argument of globalAlignment acts as tag argument that specifies the algorithm for computing a global alignment. In this example two algorithms are available: NeedlemanWunsch and Hirschberg. More tags of this kind supported by SeqAn are listed in Table 15 on page 116.

4.6.3 Defaults and Shortcuts

There are several ways to further simplify the use of SeqAn. One possibility is to define default arguments for template parameters. For example, one can write String<char> instead of String<char, Alloc<void> > in SeqAn, since the specialization Alloc is the default; see Section 7.1.

Moreover, SeqAn defines several *shortcuts* for frequently used classes. For example, we defined the type DnaString as a shortcut for String<Dna> and DnaIterator for the iterator Iterator<DnaString>::Type of DnaString. This way it is possible to program basic tasks in SeqAn even without explicitly defining any template arguments.

```
struct NeedlemanWunsch;
struct Hirschberg;

template <typename TAlignment, typename TScoring>
void globalAlignment(TAlignment & ali,
                     TScoring const & scoring,
                     NeedlemanWunsch)
{
   //Needleman-Wunsch algorithm
}

template <typename TAlignment, typename TScoring>
void globalAlignment(TAlignment & ali,
                     TScoring const & scoring,
                     Hirschberg)
{
   //Hirschberg's algorithm
}
```

Listing 6: **Tag Dispatching Example.**

Chapter 5

The Design in Examples

The examples in this chapter will demonstrate the interplay of the programming techniques that we described in Chapter 4.

5.1 Example 1: Value Counting

In this example, we want to implement a generic algorithm that counts for each value in the alphabet how often it occurs in a given string. Algorithm 1 shows the general idea of this algorithm. The implementation should at least support the following kinds of

$\triangleright$ COUNTVALUES $(a_1 \ldots a_m)$
1   $counter[c] \leftarrow 0$ for each $c \in \Sigma$
2   **for** $i \leftarrow 1$ **to** m **do**
3   $\llcorner$ $counter[a_i] \leftarrow counter[a_i] + 1$
4   report $counter$

Algorithm 1: **Algorithm for Counting String Values**. The algorithm counts for each value c of the alphabet Σ the number of occurrences of c in the string $a_1 \ldots a_m$.

string for arbitrary value types:

- Instances of SeqAn string classes `String`.

- C++ standard strings `basic_string`.

- Zero-terminated `char` arrays (C-style strings).

We will need the following functions and metafunctions: The metafunctions `Value` and `ValueSize` to determine the value type and the number of different values this type can get, the function `length` that returns the length of the string, and the function `value` for accessing the string at a given position. Note that all these functions and metafunctions are already defined in SeqAn; we will discuss how.

5.1.1   The Metafunction `Value`

The metafunction `Value` determines the *value type* of a container. For SeqAn strings, the value type is the first template argument, so we define:

```
template <typename T> struct Value;

template <typename TValue, typename TSpec>
struct Value < String<TValue, TSpec> >
{
    typedef TValue Type;
};
```

The class `basic_string` of the C++ standard library has three template arguments, and it defines the member template `value_type`, so we define a *shim* for accessing its value type as follows:

```
template <typename TChar, typename TTraits, typename TAlloc>
struct Value < basic_string<TChar, TTraits, TAlloc> >
{
    typedef basic_string<TChar, TTraits, TAlloc> TString;
    typedef typename TString::value_type Type;
};
```

A metafunction `Value` for arrays was already described in Listing 4 at page 43. We define specializations both for arrays and pointers:

```
template <typename T, size_t I>
struct Value < T [I] >
{
    typedef T Type;
};
template <typename T>
struct Value < T * >
{
    typedef T Type;
};
```

Moreover, to implement `Value` also for the const versions of these types, we specify the following rule that delegates the metafunction call to the non-const version:

```
template <typename T>
struct Value < T const >
{
    typedef typename Value<T>::Type const Type;
};
```

5.1.2 The Metafunction `ValueSize`

The metafunction `ValueSize` returns the number of different values a variable of a given type `T` can get. The default implementation uses the number of bits that are needed to store a value of type T: A type that takes n bits may store at most 2^n different values.

```
template <typename T>
struct ValueSize
{
    enum { VALUE = 1 << (sizeof(T) * 8) };
};
```

Here we use an `enum` declaration; alternatively we could also define a static member constant. Note that this implementation works on 32-bit machines only for types T with `sizeof(T)`< 4;

however, this is no serious restriction in our case, since the algorithm COUNTVALUES would not be appropriate anyway for larger alphabets.

For some alphabets which do not use all the bits for representing their values, SeqAn overloads `ValueSize` to define sharper bounds, e.g., for the nucleotide alphabet `Dna`:

```
template <>
struct ValueSize < Dna >
{
    enum { VALUE = 4 };
};
```

5.1.3   The Functions `length`

The implementation of `length` for SeqAn strings depends on the actual specialization of `String`; see Section 7.3. The length of the general purpose specialization `Alloc` for example results from the difference between the begin and the end of the string, which are both stored as data members in the object, so we may define:

```
template <typename TValue, typename TSpec> inline
typename Size< String< TValue, Alloc<TSpec> > const>::Type
length(String< TValue, Alloc<TSpec> > const & str)
{
    return end(str) - begin(str);
}
```

Note that the return value of `length` is determined by the metafunction `Size`. The default size type is `size_t`, which is sufficient for most applications. For standard strings we need again a *shim* function that wraps the member function `length` of `basic_string`:

```
template <typename TChar, typename TTraits, typename TAlloc>
inline
typename Size< basic_string<TChar, TTraits, TAlloc> >::Type
length(basic_string<TChar, TTraits, TAlloc> const & str)
{
    return str.length();
}
```

The length of C-style string is determined by searching its zero-termination:

```
template <typename T>
inline typename Size<T *>::Type
length(T * str)
{
    if (!str) return 0;
    T * it = str;
    T zero = T();
    while ( *it != zero) ++it;
    return it - str;
}
```

A *zero* is created by calling the default constructor of T. The length of a null pointer is defined to be 0.

5.1.4 The Functions value

Since all kinds of string that we consider here support the subscript operator [ ] for accessing their values, we get by with a single default implementation of value:

```
template <typename TString, typename TPosition>
inline typename Value<TString>::Type
value(TString * str,
      TPosition pos)
{
    return str[pos];
}
```

Note that a class that supports the subscript operator always implements a member function *operator [ ]*, so in order to avoid the application of member functions (see Section 4.4), generic algorithms should always use the global function `value` instead of square brackets.

5.1.5 The Generic Algorithm `countValues`

Now we have all the building blocks to implement COUNTVALUES. The result is the generic algorithm shown in Listing 7. Here we used the function `ordValue` that transforms a value of type `TValue` into `unsigned int` according to the *ord* function in Section 6.4, which maps the letters in the alphabet to numbers between 0 and the size of the alphabet -1.

```
template <typename TString>
void countValues(TString const & str)
{
    typedef typename Value<TString>::Type TValue;
    unsigned int const alphabet_size = ValueSize<TValue>::VALUE;
    unsigned int counter[alphabet_size];
    for (unsigned int i = 0; i < alphabet_size; ++i)
    {
        counter[i] = 0;
    }

    for (unsigned int i = 0; i < length(str); ++i)
    {
        TValue c = value(str, i);
        counter[ordValue(c)] += 1;
    }

    /* report counter */
}
```

Listing 7: **Generic Algorithm for Counting String Values**.

The function `countValues` can be used for all strings that support

Value, length, and value, and for all value types that support ValueSize. These functions and metafunctions may be defined for all kinds of strings and all reasonable value types, so COUNT-VALUES has potentially a very large area of application, and it is applicable to string types of different libraries, like SeqAn and the C++ standard library, as well as to built-in C-style strings. Thus we call this kind of programming *library-spanning programming*.

5.2 Example 2: Locality-Sensitive Hashing

In Section 11.2.1 we will propose the class Shape for storing a *(gapped) shape*, which is an ordered set $s = \langle s_1, \ldots, s_q \rangle$ of integers $s_1 = 1 < s_2 < \cdots < s_q$. The subsequence $a_{i+s_1} a_{i+s_2} \ldots a_{i+s_q}$ of a string $a = a_1 \ldots a_n$ is called the *(gapped) q-gram* of a at position $0 \le i \le n - s_q$. For a q-gram $b_1 \ldots b_q$, we define the hash value

$$hash(b_1 \ldots b_q) = \sum_{i=1}^{q} ord(b_i)|\Sigma|^{q-i}$$

(see Figure 10), where *ord* returns for each value of the alphabet Σ a unique integer $\in \{0, \ldots, |\Sigma| - 1\}$; see Section 6.4.

A typical task in Bioinformatics is to compute the hash values for all q-grams of a given string, e.g., for building up a (gapped) q-gram index (Section 11.2), or to apply *locality-sensitive hashing* (Indyk and Motwani 1998) for motif finding (Section 10.3.1). Listing 8 shows a generic algorithm that iterates through str and computes at each position the hash value by calling the function hash.

There are several ways for storing shapes of different kinds; see Table 28 on page 188. We will now discuss how these shape classes could be implemented in SeqAn.

5.2.1 The Base Class Shape

We decide to implement all shapes in SeqAn as refinements of the class Shape. Each shape has to know the alphabet Σ, so we specify

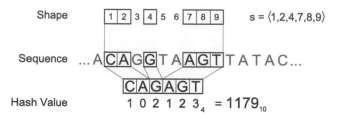

Figure 10: **Locality-Sensitive Hashing.** The example shows the application of the gapped shape $s = \langle 1, 2, 4, 7, 8, 9 \rangle$. The hash value of "CAGAGT" is 1179.

```
template <typename TShape, typename TString>
void hashAll(TShape & shape,
             TString & str)
{
    typedef typename Iterator<TString>::Type TIterator;

    TIterator it = begin(str);
    TIterator it_end = end(str) - span(shape);

    while (it != it_end)
    {
        unsigned int hash_value = hash(shape, it);
        /* do some things with the hash value */
        ++it;
    }
}
```

Listing 8: **Generic Algorithm for Computing all q-Gram Hash Values.** The function span applied to the shape $s = \langle s_1, \ldots, s_q \rangle$ returns $s_q - 1$.

this *value type* in the first template parameter of **Shape**. The actual specialization is selected in the second template parameter **TSpec**:

```
template <typename TValue, typename TSpec = SimpleShape>
class Shape;
```

The default specialization is **SimpleShape**. We will define it in Section 5.2.3. Note that there is no default implementation of Shape, i.e., all shapes classes are defined as specializations.

5.2.2 Generic Gapped Shapes

The most straightforward implementation of a generic shape $s = \langle s_1, \ldots, s_q \rangle$ stores this sequence in a data member. We use the specialization **GappedShape** of Shape for this variant.

```
template <typename TSpec = void>
struct GappedShape;

template <typename TValue, typename TSpec>
class Shape< TValue, GappedShape<TSpec> >
{
public:
    unsigned span;
    String<unsigned int> diffs;
};
```

As a shortcut for this specialization we define:

```
typedef GappedShape<> GenericShape;
```

Since it always holds that $s_1 = 1$, we need to store only $q - 1$ differences $d_i = s_{i+1} - s_i$ in the container **diffs**. Moreover, we store $s_q - 1$ in the member variable **span**, which can be retrieved by calling the function of the same name:

```
template <typename TValue, typename TSpec>
inline unsigned int
span(Shape< TValue, TSpec > const & shape)
{
    return shape.span;
};
```

Note that we define **span** in a way that it is also applicable for other specializations of **Shape**. The function **hash** can be implemented as follows:

```
template <typename TValue, typename TSpec, typename TIterator>
inline unsigned int
hash(Shape< TValue, GappedShape<TSpec> > const & shape,
     TIterator it)
{
    unsigned int val = *it;
    for (unsigned int i = 0; i < length(shape.diffs); ++i)
    {
        it += shape.diffs[i];
        val = val * ValueSize<TValue>::VALUE + *it;
    }
    return val;
};
```

5.2.3   Ungapped Shapes

The most frequently used shapes are *ungapped*, i.e., shapes $s = \langle 1, 2, \ldots, q \rangle$. Ungapped shapes can be stored much simpler than gapped shapes:

```
template <typename TValue>
class Shape< TValue, SimpleShape >
{
public:
    unsigned int span;
};
```

That means we need not to store values s_i or d_i but only the *length* q of the shape. If we know q at compile time, then we can specify it in a template parameter and define **span** as a static member:

```
template <unsigned int q = 0>
struct UngappedShape<q>;

template <typename TValue, unsigned int q>
class Shape< TValue, UngappedShape<q> >
{
public:
    static unsigned int const span = q;
};
```

We call this a *fixed shape,* since for these shapes the span q cannot be changed at run time. Since both variants of ungapped shapes are very similar (and we do not need shapes for $q = 0$) we define `SimpleShape` to be a sub-specialization of `UngappedShape` as follows:

```
typedef UngappedShape<0> SimpleShape;
```

This allows us to define functions for both kinds of ungapped shapes at once.

Ungapped shapes have the advantage that the hash value of the i-th q-gram can also be computed *incrementally* in constant type from the hash value of the $i-1$-th q-gram according to the formula:

$$hash(a_{i+1} \ldots a_{i+q}) = hash(a_i \ldots a_{i+q-1})q - a_i |\Sigma|^q + a_{i+q}$$

So we define a function `hashNext` that computes the next hash value, given the previous hash value `prev`:[1]

---

[1]Note that `hashNext` in SeqAn does not get `prev` as function argument, since the previous hash value is stored in the shape.

```
template <typename TValue, unsigned int q, typename TIterator>
inline unsigned int
hashNext(Shape< TValue, UngappedShape<q> > const & shape,
         TIterator it,
         unsigned int prev)
{
   unsigned int val = prev * ValueSize<TValue>::VALUE
                      - *it * shape.fac
                      + *(it + shape.span);
   return val;
};
```

In the above code we store the value $|\Sigma|^q$ in the member variable `fac`. In the case of fixed shapes this member variable could be a static member constant, so the compiler can apply additional optimizations which makes fixed shapes faster than shapes of variable length q.

Using `hashNext`, we can define a specialization of `hashAll` for ungapped shapes (Listing 9) that has a higher performance than the generic version in Listing 8.

5.2.4   Hardwired Shapes

We argued in the last section that fixed shapes can be faster than variable shapes, because a shape that is already defined at compile time is better optimized. Therefore we define a specialization `HardwiredShape` of `GappedShape` which encodes a (gapped) shape within template parameters:

```
template <int d1, int d2, int d3, int d4, ...>
struct HardwiredShape;
```

For this shape type the function `hash` can be computed by recursive C++ templates, which in effect cause a loop unrolling during the compilation. In practice, a *hardwired shape* achieves a much better performance than the generic gapped shape class from Section 5.2.2.

```
template <typename TValue, unsigned int q, typename TString>
void hashAll(Shape< TValue, UngappedShape<q> > & shape,
             TString & str)
{
  typedef typename Iterator<TString>::Type TIterator;

  TIterator it = begin(str);
  TIterator it_end = end(str) - span(shape);

  unsigned int hash_value = hash(shape, it);
  /* do some things with the hash value */

  while (++it != it_end)
  {
     unsigned int hash_value = hashNext(shape, it, hash_value);
     /* do some things with the hash value */
  }
}
```

Listing 9: **Special Algorithm for Computing all Hash Values of Ungapped *q*-Grams**. The first hash value is computed by hash, and the rest incrementally from the previous value by hashNext.

5.2.5   Conclusion

Figure 11 shows the hierarchy of specializations for Shape. The *left branch* Shape ← Shape<GappedShape> ← Shape<GappedShape<Hardwiredshape> > gives an example for the progressive specialization that we described in Section 4.3.1, where the *derived* class determines the TSpec slot of its *base* class. In the *right branch,* the derivation Shape<UngappedShape> ← Shape<SimpleShape> demonstrates that *template subclassing* is also capable of other kinds of class derivation: The shape class Shape<SimpleShape> is created by defining a template specialization of Shape<UngappedShape<q> > for q = 0.

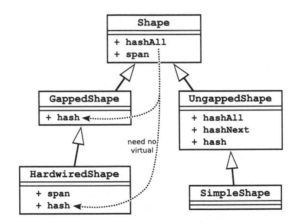

Figure 11: **Specialization Hierarchy of Shape.** The dotted pointer shows that hashAll calls the hash functions from descendant specializations. This is done without the need of *virtual* since *template subclassing* relies on *static function binding,* i.e., it is known at compile time which function is actually called.

Each specialization of Shape has its own purpose: If we want to define the actual shape at run time, then we need GenericShape or SimpleShape instead of their faster *fixed* variants HardwiredShape and UngappedShape; see Figure 12. For ungapped shapes, we better use the specializations SimpleShape or UngappedShape instead of the much slower alternatives

GenericShape or HardwiredShape.

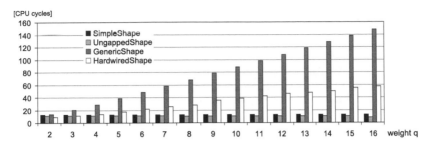

Figure 12: **Runtimes for *q*-gram Hashing.** Average runtimes for computing a hash value of an ungapped *q*-gram, where SimpleShape and UngappedShape use the function hashNext, and GenericShape and HardwiredShape the function hash. Alphabet size $|\Sigma| = 4$. The compiler optimized *fixed* versions UngappedShape and HardwiredShape take only about 80% and 40% of the time of their generic counterparts SimpleShape and GenericShape, respectively.

Part II

Library Contents

Part II gives a detailed overview of the main contents of SeqAn from the algorithmic point of view. Chapter 6 explains basic functionality of the library. Sequence data structures like strings, segments or string sets are discussed in Chapter 7, gapped sequences and sequence alignments in Chapter 8, algorithms for searching patterns or finding motifs in sequences are proposed in Chapters 9 and 10. The topic of Chapter 11 is string indices, and finally, Chapter 12 proposes the graph data structures and algorithms available in SeqAn. The string indices (Chapter 11) are in large part the work of David Weese, the graph library (Chapter 12) was implemented mainly by Tobias Rausch.

Chapter 6

Basics

In this chapter we describe basic functionality provided by SeqAn and we introduce some fundamental concepts that we will need in the following chapters. We start with the concept of *containers* of *values* in Section 6.1. The next Section 6.2 concerns memory allocation, and in Section 6.3 we explain the idea of *move* operations. The *alphabet* types provided by SeqAn are introduced in Section 6.4, and *iterators* in Section 6.5. Section 6.6 is about the *conversion* of types, and finally Section 7.11 describes the file input/output functionality in SeqAn.

6.1   Containers and Values

A *container* is a data structure that has the purpose to store *values*, i.e., objects of another data type. For example, a data structure `String` that stores the string `"ACME"` would contain the values 'A', 'C', 'M', and 'E'. Typically, all values stored in a container have the same type, we call it the *value type* of the container. The metafunction `Value` determines the value type for a given container type.

A *pseudo container* is a data structure that in fact does not store instances of the value type, but merely offers the same interface as a *real* container. Saving memory is the main reason for using pseudo containers: For example a pseudo container `vector<bool>` could store the information about its content in a bit field instead of storing individual `bool` objects in a vector. By doing this, the container will take only one bit per value instead of one byte per value and thus save memory.

The interface of containers does not depend on the way they store the information about their values. This, however, raises questions concerning the value access. A very intuitive way of accessing the values within a container is a function that returns references. A *reference* behaves like the object it refers to but has a different type. This holds in particular for C++ reference types (C++ Standard 8.3.2), e.g., `int&` that is a data type for storing references to `int` variables. It is also possible to design *proxy classes* that serve as references. Proxy classes are necessary if an access function is applied to a pseudo container, because pseudo containers do not store actual value objects, hence access functions cannot return C++ references, so the references must be emulated by a proxy class. Unfortunately, it is not possible in C++ to define proxy classes that behave correctly as references in all circumstances. For example, proxy objects usually do not fit into the same template subclassing hierarchy (Section 4.3) as the types they refer to, so different function overloads may be called if we use proxy objects instead of the values themselves as function arguments.

An alternative way to access values within a container are *get*-functions like `getValue` that either return a C++ reference to the value or, in case of a pseudo container, a temporary copy of the value. The type returned by a `getValue` can be determined by the metafunction `GetValue`, and the reference type by `Reference`. For example, `GetValue<vector<bool> >` returns `bool` and `Reference<vector<bool> >` a proxy class instead of `bool&` because `vector<bool>` is a pseudo container.

6.2   Memory Allocation

Controlling memory allocation is one of the big advantages of C++ compared to other programming languages as for example Java. Depending on the size of objects and the pattern they are allocated during the program execution, certain memory allocation strategies have advantages compared to others. SeqAn supports a variety of memory allocation strategies.

The two functions `allocate` and `deallocate` are used in SeqAn to allocate and deallocate dynamic memory (C++ Standard 3.7.3). Both functions take an allocator as an argument. An *allocator* is an object that is thought to be *responsible* for allocated memory. The default implementations of `allocate` and `deallocate` completely ignore the allocator but simply call the basic C++ operators `new` and `delete`. Although in principle every kind of object can be used as allocator, typically the object that stores the pointer to the allocated memory is used as allocator. For example, if memory is allocated for an alloc string (see Section 7.3.1), this string itself acts as allocator. A memory block should be deallocated using the same allocator object as it was allocated for.

`SimpleAlloc`	General purpose allocator.
`SinglePool`	Allocator that pools memory blocks of specific size. Blocks of different sizes are not pooled.
`ClassPool`	Allocator that pools memory blocks for a specific class. The underlying functionality is the same as for `SinglePool`.
`MultiPool`	Allocator that pools memory blocks. Only blocks up to a certain size are pooled. The user can specify the size limit in a template argument.
`ChunkPool`	Allocator that pools one or more consecutive memory blocks of a specific size.

Table 1: **Allocators.** These specializations of `Allocator` support the `clear` function.

The function `allocate` has an optional argument to specify the intended *allocator usage* for the requested memory. The user can thereby specialize `allocate` for different allocator applications. For example, the tag `TagAllocateTemp` specifies that the memory will only be used temporarily, whereas `TagAllocateStorage` indicates that the memory will be used in the long run for storing values of a container.

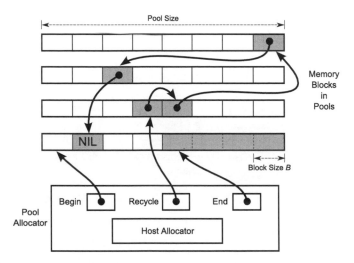

Figure 13: **Pool Allocator.** The `SinglePool` allocator optimizes the allocation of memory blocks of a certain size B that can be specified in a template argument. A host allocator – by default a `SimpleAlloc` allocator – is used to allocate the pools and requested memory blocks of size different than B. Unused memory blocks are displayed in gray. Released blocks are stored in a linked list starting with the pointer `Recycle`. If a new memory block is requested from the pool, then it is taken from the beginning of this list or, if the list is empty, the block at the end position is used.

SeqAn also offers more complex allocators which support the function `clear`. This function deallocates at once all memory blocks that were previously allocated (see for example Listing 10). The library predefines some allocator specializations for different uses (see Table 1). Most of these allocators are pool allocators. A *pool allocator* implements its own memory management: It reserves storage for multiple memory blocks at a time and recycles deallocated blocks; see Figure 13. This reduces the number of expensive `new` and `delete` calls and speeds up the allocation and deallocation; see Figure 14 for timings.

```
Allocator<MultiPool< > > mpa;

char * buf;
allocate(mpa, buf, 1000, TagAllocateTemp());
//now buf points to a char array

clear(mpa);
//all memory was deallocated
```

Listing 10: **Allocation Example.** Note that instead of `clear` we could also use `deallocate(mpa, buf, 100)` to deallocate the buffer manually.

Note that the C++ standard concept *allocator* (C++ Standard 20.1.5) differs from the SeqAn allocator concept. For example, the C++ standard requires that allocators implement several member functions, whereas the SeqAn library design avoids member functions; see Section 4.4. SeqAn offers the adaptor class `ToStdAllocator` that fulfills the allocator requirements as defined in the C++ standard and wraps the member functions `allocate` and `deallocate` to their global counterparts. One purpose of `ToStdAllocator` is to make standard containers use the SeqAn allocators for retrieving memory.

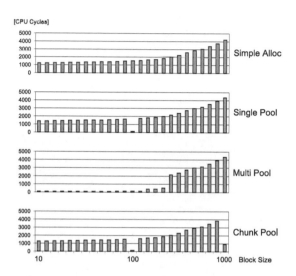

Figure 14: **Allocator Run Times.** The average time for allocating memory
   blocks of different sizes using (1) `SimpleAlloc` (2) `SinglePool<100>` (3)
   `MultiPool` and (4) `ChunkPool<100>`. The time for getting memory from
   `SimpleAlloc` reflects the expense for requesting it directly from the heap
   manager. Blocks of size 100 (in case of `SinglePool<100>`) or multiples
   of 100 (in case of `ChunkPool<100>`) are taken from the pool; `MultiPool`
   pools blocks of size ≤ 256. All other blocks are requested from the heap
   manager. The figure shows that getting a memory block from a pool
   takes approximately 8% of the time needed to allocate the same amount
   of memory from the heap manager.

6.3 Move Operations

There is often an opportunity to speed up copy operations, if the source is not needed any more after copying. Therefore, we introduce *move* operations, i.e., assignments that allow the destruction of the source. For example, if a container stores its values in a dynamically allocated memory block, a move operation may simply pass the memory block of the source object to the target. The source container will be empty after the move operation. Move operations like `moveValue` are alternatives for regular assignment functions like `assignValue`.

```
String<char> str1 = "ACGT";
String<char> str2(str1, Move());
cout << str2;            //output: "ACGT"
cout << length(str1);    //output: 0
```

Listing 11: **Move Constructor Example.**

In many cases, SeqAn also offers special constructors that apply move operations. The *move constructor* differs from the regular copy constructor in an additional tag argument `Move` (see Listing 11).

6.4 Alphabets

A value type that can take only a limited number of values is called a (finite) *alphabet* Σ. We can retrieve the number of different values of an alphabet $|\Sigma|$, the *alphabet size*, by the metafunction `ValueSize`. Another useful metafunction called *BitsPerValue* can be used to determine the number of bits needed to store a value of a given alphabet. Table 2 lists some alphabets predefined in SeqAn. Let $\Sigma = \{\sigma_0, \ldots, \sigma_{|\Sigma|-1}\}$ be an alphabet, then we denote $ord(\sigma_i) = i$. This number can be retrieved by calling the function `ordValue`. All predefined alphabets in SeqAn store their values in enumerated integers $\{0, \ldots, \mathtt{ValueSize}-1\}$, so `ordValue` is for those value types a trivial function.

Dna	Alphabet for storing nucleotides of deoxyribonucleic acid, i.e., 'A', 'C', 'G', and 'T'.
Dna5	Like Dna, but with an additional value 'N' for *unknown nucleotide*.
Rna	Alphabet for storing nucleotides of ribonucleic acid, i.e., 'A', 'C', 'G', and 'U'.
Rna5	Like Rna, but with an additional value 'N' for *unknown nucleotide*.
Iupac	Iupac code for storing nucleotides of DNA/RNA. The Iupac codes are enumerated in this order: 'U'= 0, 'T', 'A', 'W', 'C', 'Y', 'M', 'H', 'G', 'K', 'R', 'D', 'S', 'B', 'V', 'N'= 15
AminoAcid	Alphabet for storing amino acids.

Table 2: **Alphabets in** SeqAn. The listed characters are the result when a value is converted into `char`.

6.4.1   Simple Types

Containers in SeqAn are usually designed as generic data structures that can be instantiated for arbitrary value types. The value type can therefore be any user-defined class as well as a simple type. A *simple type* is a type that does not need a constructor to be created, a destructor to be destroyed, and neither a constructor nor an assignment operator to be copied.

Simple objects have the advantage that they can be moved within the computer's main memory using fast memory manipulation functions. In many cases, containers that work on simple types can therefore be implemented much faster than generic containers that must copy values one after another using the correct assignment operator or copy constructor.

POD (*plain old data*) types (C++ Standard 3.9) are simple, for example built-in types like `char` or `wchar_t`. A C++ class can also be simple even if it defines constructors, destructors or assignment operators, as long as these functions are not necessary for correctly creating, destroying, or copying instances of this class. All value types listed in Table 2 are simple.

The metafunction `IsSimple` can be used to distinguish between simple and non-simple types in metaprogramming.

6.4.2 Array Operations

In SeqAn a set of array operations serve as an abstraction layer to apply divergent handling between simple types and other kinds of types; see Table 3. For example, the general version of the function `arrayCopy` uses a loop to copy a range of objects into a target range, whereas a specialized version of `arrayCopy` for simple types applies the fast memory manipulation function `memmove` (C++ Standard 20.4.6).

6.4.3 Alphabet Modifiers

A *modifier* is a class that adapts types in a way that the adapted type is still of the same kind but shows some differences compared to the unmodified type. The alphabet expansion modifier `ModExpand` for example transforms an alphabet into another alphabet that contains an additional character, i.e., the *value size* is increased by one. For example `ModifiedAlphabet<TValue, ModExpand<'-'> >` expands the alphabet `TValue` by a gap character '-'. This alphabet is used in the context of gapped sequences and alignments; see Chapter 8. It is returned by the metafunction `GappedValueType` for value types that do not already contain a gap value.

SeqAn also offers string modifiers; see Section 7.6.

`arrayClearSpace`	Destroys a part of an array starting from the beginning and keeps the rest.
`arrayConstruct`	Construct objects in a given memory buffer.
`arrayConstructCopy`	Copy constructs an array of objects into a given memory buffer.
`arrayConstructMove`	Move constructs an array of objects into a given memory buffer.
`arrayCopy`	Copies a range of objects into another range of objects.
`arrayCopyBackward`	Copies a range of objects into another range of objects starting from the last element.
`arrayCopyForward`	Copies a range of objects into another range of objects starting from the first element.
`arrayDestruct`	Destroys an array of objects.
`arrayFill`	Assigns one object to each element of a range.
`arrayMove`	Moves a range of objects into another range of objects.
`arrayMoveBackward`	Moves a range of objects into another range of objects starting from the last element.
`arrayMoveForward`	Moves a range of objects into another range of objects starting from the first element.

Table 3: **Array Operations.**

6.5 Iterators

An *iterator* is an object that is used to browse through the values of a container. The metafunction `Iterator` can be used to determine an appropriate iterator type given a container. Figure 15 shows some examples. Some containers offer several kinds of iterators, which can be selected by an optional argument of `Iterator`. For example, the tag `Standard` can be used to get an iterator type that resembles the C++ standard random access iterator (see C++ Standard 24.1.5). The more elaborated *rooted iterator*, i.e., an iterator that *knows* its container, can be selected by specifying the `Rooted` tag.

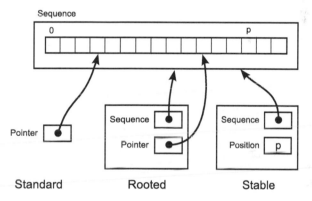

Figure 15: **Iterators for Alloc Strings.** See Section 7.3.1. The standard iterator is just a pointer to a value in the string. The rooted iterator also stores a pointer to the string itself. The stable iterator stored the position instead of a pointer to a value since pointers could be invalid when the alloc string is resized.

Rooted iterators offer some convenience for the user: They offer additional functions like `container` for determining the container on which the iterator works, and they simplify the interface for other functions like `atEnd`; see Listing 12. Moreover, rooted iterators may change the container's length or capacity, which makes it possible to implement a more intuitive variant of a **remove** algorithm (see C++ Standard 25.2.7). On the other hand, standard iterators can often be implemented simply as pointers, and in practice they are faster than rooted iterators, which typically suffer from an *abstraction penalty*; see Section 4.2. Hence, the default iterator is set to Standard for most containers. This default is defined by the metafunction DefaultIteratorSpec.

```
String<char> str = "ACME";
Iterator<String<char>, Rooted> it;       //a rooted iterator
for (it = begin(str); !atEnd(it); ++it)
{
    cout << *it;
}
```

Listing 12: **Rooted Iterator Example.** Since it is a rooted iterator, it supports the unary function atEnd that returns true if and only if the iterator points behind the end of its container. A standard iterator that does not *know* its container could not support this function.

While rooted iterators can usually be converted into standard iterators, it is not always possible to convert standard iterators back into rooted iterators, since standard iterators may lack the information about the container they work on. Therefore, many functions that return iterators like begin or end return rooted iterators instead of standard iterators; this way, they can be used to set both rooted and standard iterator variables. Alternatively it is possible to specify the returned iterator type explicitly by passing the iterator kind as a tag argument (see Listing 13). An iterator is *stable* if it stays valid even if its container is expanded, otherwise it is *unstable*. For example, the standard iterator of alloc strings (Section 7.3.1) – which is a simple pointer to a value in the string – is unstable, since during the expansion of an alloc string, all values are moved to new memory addresses. A typical implementation of stable iterators for strings stores the position instead of a pointer to the current value. The Iterator metafunction called with the Stable tag returns a type for stable iterators.

```
String<char> str = "ACME";
Iterator<String<char> >           it1 = begin(str);
//a standard iterator
Iterator<String<char>, Standard> it2 = begin(str);
//same as above
Iterator<String<char>, Rooted>    it3 = begin(str);
//a rooted iterator
Iterator<String<char>, Rooted>    it4 = begin(str, Rooted());
//same as above
```

Listing 13: **Examples for Creating Iterators.** If no iterator kind is specified, the metafunction `Iterator` assumes `Standard` and the function `begin` assumes `Rooted`. Both `it1` and `it2` are standard iterators, whereas `it3` and `it4` are rooted iterators.

6.6   Conversions

The function `convert` transforms objects from one type `TSource` into another type `TTarget` (see Listing 14). There are two possibilities for doing that: If the object can simply be reinterpreted as an object of type `TTarget`, `convert` returns a `TTarget&` referring to the original object. Otherwise, `convert` returns a temporary (C++ Standard 12.2) object of type `TTarget`. The actual return type can be determined by the metafunction `Convert`.

```
TSource obj;
Convert<TTarget, TSource>::Type obj2 = convert<TTarget>(obj);
```

Listing 14: **Value Conversion.** The program converts an object of type `TSource` into an object of type `TTarget`. `Convert<TTarget, TSource>::Type` is either `TTarget&` or `TTarget`, according to whether `obj` can be reinterpreted as `TTarget` object or not.

Chapter 7

Sequences

7.1 Strings

A *sequence* is a container that stores an ordered list of values, like nucleotides, amino acids or `char` values. Examples for sequences are "hello world" or "ACGT". The number of these values is called the *length* of the sequence. The values in a sequence are ordered. We define $i - 1$ to be the *position* of the i-th value in a sequence, i.e., the first value in the sequence stands at position 0 and the last at position $length - 1$, as it is standard in C. We call 0 the *begin position* and the length of the sequence the *end position*. Note that the end position is not the position of the last value in the sequence but the position after the last value.

SeqAn implements several string types as specializations of the class `String`. These specializations are described in Section 7.3. There is another class `StringSet` that offers an implementation of *strings of strings*, i.e., strings that use again strings as value types. `StringSet` will be described in Section 7.9.

All string classes in SeqAn (with the exception of the external string; see Section 7.3.5) are designed as generic data structures that can be instantiated for all kinds of value, i.e., for both *simple* and *non-simple* value types (Section 6.4.1). Typically, the value type is qualified as the *first* template argument in angled brackets:

```
String<AminoAcid> myProteine;
```

SeqAn predefines shortcuts for some usual value types, so we can also write:

```
Peptide myProteine; //same as above
```

The user can specify the specialization of string (Table 6 on page 85) that should be used in the optional *second* template argument of `String`, for example:

```
String<char, Array<100> > myArrayString;
//string with maximum length 100
String<Dna, Packed<> > myPackedString;
//string that takes only 2 bits per nucleotide
```

A sequence is *contiguous* if it stores its values consecutively in a single memory block. Examples for contiguous strings are the standard library `basic_string` (see C++ Standard 21.3) and simple `char` arrays. Applied to a sequence type T, the metafunction `IsContiguous` returns `True` if T is contiguous, otherwise `False`. SeqAn offers many functions and operators for initializing, converting, manipulating, and printing strings; see Table 4. An example for string manipulation is given in Listing 15.

7.2   Overflow Strategies

Some sequence types reserve space of storing values in advance. The number of values for which a sequence has reserved space is called the *capacity* of this sequence. The capacity is therefore an upper bound for the length of a sequence. A sequence is called expandable, if its capacity can be changed. All string classes in SeqAn– except of the *array string* (see Section 7.3.2) – are expandable. Changing the capacity can take much time, e.g., expanding an alloc string (see Section 7.3.1) necessitates to copy all values of this string into a new memory block.

There are numerous functions in SeqAn that can change the length of a sequence. If the current capacity of a sequence is

Metafunctions	
Value	The *value type* of the sequence.
Iterator	The type of *iterators* for browsing the sequence (see Section 6.5).
Size	The type for storing the *length* or *capacity* of a sequence. This is `size_t` in most cases.
Position	An unsigned integral type to store the *position* of a value within the sequence.
Infix	A segment type that can represent an *infix* of the sequence.
Prefix	A segment type that can represent a *prefix* of the sequence.
Suffix	A segment type that can represent a *suffix* of the sequence.
Functions	
assign	Copies a sequence to another sequence. The same as the assignment `operator =`.
append	Appends a sequence to another sequence. The same as the `operator +=`.
replace	Replaces a part of the sequence with another sequence.
value	Returns a *reference* to a value in the sequence at a given position. The same as the subscription operator `[ ]`.
begin	Returns an *iterator* to the begin of the sequence.
end	Returns an *iterator* to the end of the sequence.
length	The number of values stored in the sequence.
empty	Returns `true` if the sequence is empty, i.e., its length is 0, otherwise `false`.
clear	Sets the length of the sequence to 0.
capacity	The maximal length of the sequence. Can be changed using `reserve`.
reserve	Changes the *capacity* of the sequence.

Table 4: **Common Functions and Metafunctions for Sequences.**

```
#include <iostream>
#include <seqan/sequence.h>
#include <seqan/file.h>

using namespace seqan;
using namespace std;
int main()
{
  String<AminoAcid> prot = "anypeptide";
  cout << length(prot) << endl;
  //output: 10

  prot += "anewend";
  cout << prot << endl;
  //ouput: "ANYPEPTIDEANEWEND"

  return 0;
}
```

Listing 15: **Example Program for Strings**.

Exact	Expand the sequence exactly as far as needed. The capacity is only changed if the current capacity is not large enough.
Generous	Whenever the capacity is exceeded, the new capacity is chosen somewhat larger than currently needed. This way, the number of capacity changes is limited in a way that resizing the sequence only takes amortized constant time.
Limit	Instead of changing the capacity, the contents are limited to current capacity. All values that exceed the capacity are lost.
Insist	No capacity check is performed, so the user has to ensure that the container's capacity is large enough.

Table 5: **Overflow Strategies**.

exceeded by changing the length, we say that the sequence over-flows. The *overflow strategy* (see Table 5) determines the behavior of a sequence in the case of an overflow. The user can specify the overflow strategy by applying a *switch argument*. Otherwise the overflow strategy is determined depending on the sequence class: For functions that are used to *explicitly* change a sequence's length (like `resize` or `fill`) or capacity (`reserve`), the metafunction `DefaultOverflowExplicit` specifies the default overflow strategy. Functions like `appendValue` or `replace` that primarily serve other needs than changing lengths or capacities may also cause an overflow *implicitly*. For these functions, the metafunction `DefaultOverflowImplicit` is used to determine the default overflow strategy. For example, the alloc string uses `Exact` as explicit and `Generous` as implicit default expansion strategy. Listing 16 gives an example of the effect of overflow strategies. The overflow

```
String<char> str;

//default expansion strategy Exact:
resize(str, 5);
//now the capacity of str is 5

//use expansion strategy Limit:
assign(str, "abcdefghi", Limit());
//only "abcde" was assigned to str.

//default expansion strategy Generous:
append(str, "ABCDEFG");
//now str == "abcdeABCDEFG"
```

Listing 16: **Overflow Strategies Example.**

strategy `Generous` is used to achieve amortized constant costs for appending single values to a string, e.g., the alloc string. When the string is expanded, the function `computeGenerousCapacity` is called to compute a new capacity for this string. The default implementation – which can be overwritten by the user – reserves

50% extra space for storing values. This additional memory is used to store values that are appended afterwards. One can easily show that the number of expansions of a string is logarithmic in the number of value appends, and that each value in the string is moved at most three times on average.

7.3   String Specializations

We will now describe the different specializations of the class `String` (see Table 6 for an overview and Figure 16 for timings).

7.3.1   Alloc Strings

SeqAn offers two contiguous string specializations: `Alloc` and `Array`. `Alloc` strings use dynamic memory (C++ Standard 3.7.3) for storing values. Expanding the string means therefore that we need to move all values into a new allocated larger memory block. That in turn makes most iterators unstable (see Section 6.5). The amortized costs for appending a value (e.g., using `appendValue`) is constant if the overflow strategy *Generous* is used (see Section 7.2).

Since alloc strings are a good choice for most applications, `Alloc` is the default string specialization. See Figure 17 for an illustration.

7.3.2   Array Strings

`Array` strings store values in an array data member. This array has a fixed size, which is specified by a template argument. The advantage of array strings is that no expenses are incurred for allocating dynamic memory if the string is created with static or automatic storage duration (C++ Standard 3.7.2), i.e., the string is stored on the call stack at compile time. This can also speed up the value access, since the call stack is a frequently used part of the memory and has therefore a good chance to stay in the cache. On the other hand, the finite size of the call stack limits the capacity of

`Alloc`	The default string implementation that can be used for general purposes. The values are stored in a contiguous memory block on the heap. Changing the capacity can be very costly since all values must be copied into a new memory block.
`Array`	A fast but non-expandable string that stores its values in a member variable and that is therefore best suited for holding small temporary sequences.
`Block`	A string that stores its content in blocks, so that the capacity of the string can quickly be increased without copying existing values. Though iteration and random access to values are slightly slower than for alloc strings, block strings are a good choice for growing strings and stacks.
`Packed`	A string that stores as many values in one machine word as possible and that is therefore suitable for storing very large strings in memory. Since each value access takes some bit operations, packed strings are in general slower than other in memory strings.
`External`	A string that stores the values in secondary memory (e.g., a hard disk). Only parts of the string are loaded into main memory whenever needed. This way, the total string length is not limited by the machine's physical memory.

Table 6: **String Specializations.**

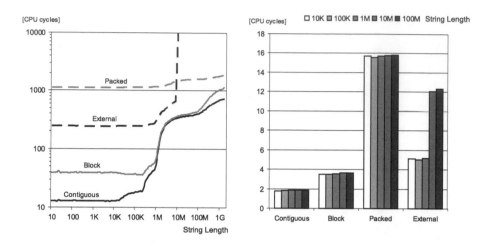

Figure 16: **String Value Accessing Run Times. Left**: Run times for copying a random value into a random place depending on the length of the strings. **Right**: Run times for iterating a string and moving the whole string one value further.

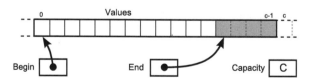

Figure 17: **Contiguous String.** The figure shows an alloc string. The values are stored in a single contiguous piece of memory. The string also stores the capacity of the storage and the begin and end of the currently used part.

the string. A typical application for the array strings is to provide quickly limited storage for sequences.

7.3.3   Block Strings

A `Block` string stores values in a set of fixed-size memory blocks. The location of these blocks is stored in a directory. Block strings are expanded by adding new memory blocks. The advantage – compared to contiguous strings – is that this can be done without moving values in memory. Block string iterators are therefore always stable (see Section 6.5), and it is uncritical to store pointers to values that are stored in a block string. Random access to a value at a given position in the block string is done in four steps: (1) Determine the number of the block the value is stored, (2) look up in the directory the location of the block, (3) determine the offset at which the value is stored within the block, (4) access the value. If the block size is set to a power of two, step (1) take only one **shift**- and step (3) one **and**-operation. Nevertheless, random accesses to values in block strings are up to three times slower than random accesses to values in contiguous strings, and iterating over a block string takes about two times longer than iterating over a contiguous string (see Figure 16).

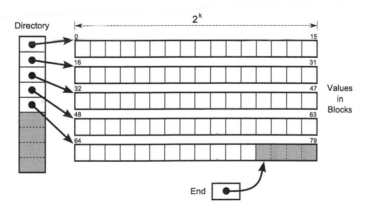

Figure 18: **Block String.** The values are stored in a set of blocks, each of the same size that is a power of two. A directory sequence stores pointers to these blocks. All blocks except the last one are completely filled.

The block string is optimized for appending and removing single values at the end of the string. It supports the functions **push** – a synonym for **appendValue** – and **pop**, and it is therefore best suited to be used as stack.

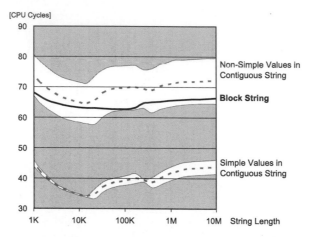

Figure 19: **Append Value Times.** The average time needed to append a single value to a string depending on the string length. Appending a single value to a contiguous string could be very expensive, if this causes an expansion of the string buffer, which means that the complete string must be copied. The two displayed corridors give the upper and lower bound for appending (1) a simple type value or (2) a non-simple type value to a contiguous string, where the **sizeof** of a single value is 1. The upper bound is reached, if the string was expanded during the last appending, and the lower bound, if the next appending would cause a buffer expansion. Block strings need no buffer expansions, so the time for appending a value is less dependent from the kind of object stored in it.

7.3.4   Packed Strings

Objects in C++ have at least a size of one *byte* (C++ Standard 1.7), hence each value takes at least eight *bits*. But this is more than actually needed for some value types. For example, the alphabet **Dna** has only four letters 'A', 'C', 'G', and 'T', so that a **Dna** value can be encoded in only two bits. The metafunction **BitsPerValue** returns the number of bits needed to store a value.

The `Packed` string stores the values *packed*, i.e., each value takes only the minimal number of bytes. For example, the packed string compresses a `Dna` sequence down to a quarter of its *unpacked* size.

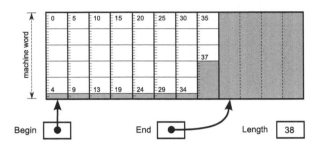

Figure 20: **Packed String.** The string stores as many values per machine word as possible. In this example, each value takes six bits. The packed string stores five instead of four values per 32-bit machine word, only two bits per machine word are wasted.

However, the handling of packed strings is slower than for all other in-memory string types in SeqAn. In practice, random accesses in packed strings are up to two orders of magnitude slower than random accesses in contiguous strings (see Figure 16). This difference has three reasons: (1) Accessing a value in a packed string is much more complicated than accessing a value in a contiguous string, because each access takes multiple operations to filter out the relevant bits, (2) the high complexity of access operations obstruct an efficient optimization by the compiler, and (3) since packed strings are suitable to handle very long sequences, SeqAn packed string uses 64-bit position types, but this slows down random accesses on 32-bit machines by a factor of about two. The packed string's iterator was optimized to speed up accesses, but an iteration still takes eight times longer than iterating an unpacked contiguous string (see Figure 16). For that reason, the application of packed strings is only advisable if the handled sequences are too long to be stored in main memory.

7.3.5 External Strings

The External string stores its values on external memory, i.e., in a file, with the effect that the main memory size does not limit the sequence length any more. In particular, external strings can be larger than 4 GB even on 32-bit machines, where we then need 64-bit words to store a positions of a value within the string. The file is organized into fixed length blocks, and only some of them are cached in main memory. Both block length and number of cached blocks can be specified in template arguments. When the user accesses a value of an uncached block, the block is loaded into memory, and in return, the least recently used block in the cache is written back to the file. During an iteration, the external string's iterator prefetches asynchronously the next-in-line memory block. This *trick* speeds up the sequential iteration, but random accesses to values in external strings are very slow; see Figure 16.

7.4 Sequence Adaptors

SeqAn also implements complete string interface adaptors – both functions and metafunctions – for data types that are not part of SeqAn. This way, these data types can be accessed in the same way as SeqAn sequences, i.e., string algorithms in SeqAn can be applied to these data types. There are three adaptions:

(1) For *zero terminated* **char** *arrays*, also known as *C-style strings,* the *classical* way for storing strings in the programming language C. For example, the **length** function for C-style strings calls the standard library function **strlen**. The interface applies to arrays of **char** or **char_t** and to **char *** and **wchar_t *** pointers:

```
char str[] = "this is a char array";
cout << length(str);          //output: 20
```

Unfortunately, it is not possible to distinguish between C-style strings and `char` pointers, which could also be iterators for C-style strings or other string classes, so if the user calls for example `append` to attach a `char *` to a string `str`, then we could either append a sequence of `char` or a single `char *`. Note that it is not possible to decide this just regarding the value type of `str`, since this could be any type into which either `char` or `char *` can be converted.

Another limitation of C-style strings is that we cannot define all common operators like `operator =` for built-in types like `char *`.

(2) For the standard library class `basic_string`, that is widely used in C++ programs. For example, the `length` function for `basic_string` string calls the member function `length`:

```
std::basic_string str = "a standard library string";
cout << length(str);            //output: 25
```

(3) Finally, there is also a generic sequence interface that applies to any data type (if there is no further implementation) in a way, that an arbitrary object is regarded as a sequence of these objects of length 1:

```
MyClass obj;
cout << length(obj);            //output: 1
```

7.5 Iterating Sequences

Iterators are objects that are used to scan through containers like strings or segments (Section 6.5). Listing 17 shows an example. An iterator always points to one value in the container. The function `value` (which does the same as the operator `*`) can be used to access this value. Functions like `goNext` or `goPrevious` (which do the same as ++ and --, respectively) can be used to move the iterator to other values within the container.

```
String<char> str = "acgt";
typedef Iterator<String<char> >::Type TIterator;
for (TIterator it = begin(str); it != end(str); ++it)
{
    cout << value(it);
}
//output: "start_overwrite_end";
```

Listing 17: **Iterating a String.**

The functions `begin` and `end` applied to a container return iterators to the begin and the end of the container. Note that, similar to C++ standard library iterators, the iterator returned by end does not point to the last value of the container but to the value that would come next. So if a string `s` is empty, then `end(s) == begin(s)`.

7.6   Sequence Modifiers

SeqAn supports several *modifiers* (see Section 6.4.3) for strings that allow a different *view* to a given string; see Table 7. Modifiers for strings are implemented in specializations of the class `ModifierString`. The required space of a modifier is constant, i.e., it does not depend on the length of the sequence. Each modifier exports a complete string interface, including an appropriate iterator and segment data types (Section 7.7).

One specialization of `ModifiedString` is the `ModView` modifier: Assume we need all characters of `myString` to be in uppercase without copying `myString`. We first create a *functor*, i.e., an STL unary function, which converts a character to uppercase:

ModReverse	The reverse $a_n \ldots a_1$ of the string $a_1 \ldots a_n$.
ModView	Transforms the values of a string $a_1 \ldots a_n$ using a custom functional. The type of the functional is specified as template argument of `ModView`. SeqAn offers the following predefined functionals:
	`ModView<FunctorConvert>`: Converts the value type.
	`ModView<FunctorLowcase>`: Converts to lowercase characters, e.g., 'A' is converted to 'a'.
	`ModView<FunctorUpcase>`: Converts to uppercase characters, e.g., 'b' is converted to 'B'.
	`ModView<FunctorComplement>`: Converts nucleotide value `Dna` or `Dna5` to their complement, e.g., 'A' is converted to 'T', 'C' to 'G', and vice versa.

Table 7: **String Modifiers.**

```
struct FunctorUpcase: unary_function<char, char>
{
    inline char operator()(char x) const
    {
        if (('a' <= x) && (x <= 'z')) return (x + ('A' - 'a'));
        else return x;
    }
};
```

This functor `FunctorUpcase` is already part of SeqAn, together with some other predefined functors. We create a `ModifiedString` specialized with `ModView<FunctorUpcase>` as follows:

```
String<char> str = "A man, a plan, a canal";
typedef ModifiedString<
            String<char>,
            ModView< FunctorUpcase<char> > > TMod;
TMod str_upper(str);
cout << str_upper << endl; //output: "A MAN, A PLAN, A CANAL"
```

Modifiers can also be nested, for example the following program shows how to get the reverse complement of a given Dna string by applying two modifiers on it:

```
String<Dna> myString = "attacgag";
typedef ModifiedString<String<Dna>, ModComplementDna> TConv;
typedef ModifiedString<TConv, ModReverse> TRevConv;
TRevConv myReverseComplement(myString);
std::cout << myReverseComplement << endl;   //prints "CTCGTAAT"
```

Since accessing a string through a modifier causes a certain overhead, it could be advisable to convert the string itself – though this has the disadvantage that the original string gets lost. SeqAn therefore offers *in-place* modifier functions reverseInPlace and convertInPlace. The following example program converts a Dna string to its reverse complement:

```
String<Dna> myString = "attacgag";
convertInPlace(myString, FunctorComplement<Dna>());
reverseInPlace(myString);
std::cout << myString << endl;       //prints "CTCGTAAT"
```

7.7   Segments

A *segment* is a contiguous part of a sequence. The sequence is called the *host* of the segment. SeqAn implements segment data types for infixes, prefixes, and suffixes: A *prefix* is a segment that starts with the first value of the host, a *suffix* is a segment that ends with the last value of the host, and an *infix* is an arbitrary segment.

The segment data structures in SeqAn are pseudo-containers: They do not store values themselves but a link to their host and the begin and end borders of the segment; see Figure 21. These borders can be set either during the construction of the segment or by functions like setBegin or setEnd. Changing the content of a segment means to change the content of its host; see Listing 18.

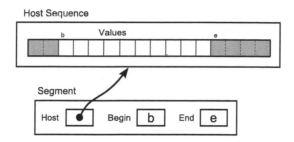

Figure 21: **Segment.** The infix segment stores a pointer to the host and the begin and end position of the subsequence.

```
String<char> str = "start_middle_end";
Infix<String<char> > inf = infix(str, 6, 12);
cout << inf;          //output: "middle"
inf = "overwrite";
cout << str;          //output: "start_overwrite_end";
prefix(str, 5) = "XYZ";
cout << str;          //output: "XYZ_overwrite_end";
```

Listing 18: **Segment Example.** This program demonstrates how the content of a string `str` can be changed by assigning new values to segments. If this effect is undesirable, one has to explicitly make a copy of the string.

The metafunctions `Infix`, `Prefix`, and `Suffix`, respectively, return for a given sequence an appropriate data type for storing the segment. The functions `infix`, `prefix`, and `suffix` create temporary segments that can directly be used to manipulate their host sequence. It is also possible to create segments of segments, but this does not introduce new types: A segment A of a segment B of a sequence S is again a segment of S. Note that changing the borders of B does not affect the borders of A.

7.8   Comparators

Two sequences can be lexicographically compared using usual operators like < or >=, for example:

```
String<char> a        = "beta";
String<char> b        = "alpha";

bool a_not_equals_b = (a != b);      //true
bool a_less_b       = (a < b);       //false
```

Each comparison involves a scan of the two sequences for searching the first mismatch between the strings. This could be expensive if the two sequences share a long common prefix. Suppose for example that we want to branch in a program depending on whether $A < B$, $A == B$, or $A > B$, for example:

```
if (A < B)      { /* code for case "A < B"  */ }
else if (A > B) { /* code for case "A > B"  */ }
else            { /* code for case "A == B" */ }
```

In this case, although only one scan would be enough to decide what case is to be applied, each operator > and < performs a new comparison. SeqAn offers lexicals to avoid such unnecessary sequence scans. A *lexical* is an object that stores the result of a comparison. Applying a lexical to the example above leads to the following code:

```
Lexical<> comp(A, B);
if (isLess(comp))          { /* code for case "A < B"  */ }
else if (isGreater(comp))  { /* code for case "A > B"  */ }
else                       { /* code for case "A == B" */ }
```

The two sequences A and B are compared during the construction of the lexical comp. The result is stored in the lexical and is accessed via the functions isLess and isGreater.

7.9 String Sets

A set of sequences can either be stored in a sequence of sequences, for example in a `String< String<char> >`, or in `StringSet`. One advantage of using `StringSet` is that it supports the function `concat` that returns a *concatenator* of all sequences in the string set. A *concatenator* is an object that represents the concatenation of a set of strings. This way it is possible to build up index data structures for multiple sequences by using the same construction methods as for single sequences (see Chapter 11). The specialization `Owner<ConcatDirect>` already stores the sequences in a concatenation. The concatenators for all other specializations of `StringSet` are *virtual* sequences, that means their interface *simulates* a concatenation of the sequences, but they do not literally concatenate the sequences into a single sequence. Hence in any case the sequences need not to be copied when a concatenator is created.

There are two kinds of `StringSet` specializations in SeqAn: `Owner` and `Dependent`; see Table 8. `Owner` string sets actually store the sequences, whereas `Dependent` string set just refer to sequences that are stored outside of the string set.

One string can be an element of several `Dependent` string sets. Typical tasks are therefore to find a specific string in a stringset, or to test whether the strings in two string sets are the same. Therefore a mechanism to identify the strings in the string set is needed, and, for performance reasons, this identification should not involve string comparisons. We solved this problem by introducing *ids*, which are by default `unsigned int` values. There are two ways for accessing the sequences in a string set: (1) the function `value` returns a reference to the sequence at a specific *position* within the sequence of sequences, and (2) `valueById` accesses a sequence given its *id*. In the case of `Owner` string sets, id and position of a string are always the same, but for `Dependent` string sets, the ids can differ from the positions. For example, if a `Dependent` string set is used to represent subsets of strings that are stored in `Owner` string sets, one can use the position of the

`Owner`	The default specialization of `StringSet`. The sequences in this string set are stored in a string of string data structure. `concat` returns a special *concatenator* object that simulates the concatenation of all these strings.
`Owner<ConcatDirect>`	The sequences are stored as parts of a long string. Since the sequences are already concatenated, `concat` just needs to return this string. The string set also stores lengths and starting positions of the strings. Inserting new strings into the set or removing strings from the set is more expensive than for the default `Owner` specialization, since this involves moving all subsequent sequences in memory.
`Dependent<Tight>`	This specialization stores sequence pointers consecutively in an array. Another array stores an id value for each sequence. That means that accessing given an id needs a search through the id array.
`Dependent<Generous>`	The sequence pointers are stored in an array at the position of their ids. If a specific id is not present, the array stores a zero at this position. The advantage of this specialization is that accessing the sequence given its id is very fast. On the other hand, accessing a sequence given its position i can be expensive, since this means we have to find the i-th non-zero value in the array of sequence pointers. The space requirements of a string set object depends on the largest id rather than the number of sequences stored in the set. This could be inefficient for string sets that store a small subset out of a large number of sequences.

Table 8: **StringSet Specializations.**

string within the `Owner` string set as id of the strings.

7.10   Sequence Conversion

A sequence of one value type can be converted into a sequence of another value type, if the two value types are convertible. SeqAn offers three different ways for sequence conversion:

(1) **Copy conversion**. The source sequence is copied into the target sequence, e.g., during construction, by assignment (operator =), or using the function `assign`.

```
String<Dna> source = "acgtgcat";
String<char> target;
assign(target, source);      //copy conversion
```

(2) **Move conversion**. In some cases, the function `move` can perform an in-place conversion. For example, if source and target sequence are `Alloc` strings (see Section 7.3.1) and if the two value types have the same size, then `move` transfers the value storage of the source to the target string. After that, all values are converted to the new value type.

```
String<Dna> source = "acgtgcat";
String<char> target;
move(target, source);           //in-place move conversion
```

(3) **Modifier conversion**. A modifier can *emulate* a sequence with a different value type instead of creating an actual target sequence; see Section 7.6.

```
String<Dna> source = "acgtgcat";
typedef Modifier<String<Dna>,
    ModView<FunctorConvert<Dna, char> > > TModifier;
TModifier target(source);    //char sequence "acgtgcat"
Value<TModifier>::Type c;    //a variable of type char
```

Raw	The default file format. `Raw` applied to sequences means that the file content is directly interpreted as a sequence. `Raw` applied for writing an alignment generates a pretty print.
Fasta	A common file format for storing sequences or alignments (Pearson and Lipman 1988). Each record consists of a single line starting with '>' that contains metadata, followed by the sequence data.
Embl	The EMBL/Swissprot file format (Stoehr and Cameron 1991) for storing sequences and complex metadata. Each metadata entry starts with a two-letter code (see EMBL User Manual 2008).
Genbank	The GenBank file format (Benson et al. 2008); an alternative notation of EMBL/Swissprot file format for sequence data.
DotDrawing	File format for graphs (see Chapter 12), write only.

Table 9: **Some File Formats.**

7.11 File Input/Output

SeqAn supports several ways for reading and writing sequences and alignments in different file formats. Table 9 shows some supported file formats. There are two ways for accessing a file in SeqAn: (1) file access functions and (2) file reader stings.

7.11.1 File Access Functions

The function **read** loads data from a file and **write** saves data to a file. Both C-style **FILE** handles or C++ stream objects can be used as files. Note that the files should always be opened in binary mode. The simplest file format is `Raw` that is used to load a file *as is* into a string or vice versa, e.g.,:

```
//loading
fstream fstrm;
fstrm.open("input.txt", ios_base::in | ios_base::binary);

CharString str;
read(fstrm, str, Raw());

fstrm.close();

//saving
FILE * cstrm = fopen("output.txt", "w");
write(cstrm, str, Raw());
fclose(fstrm);
```

In this example, the tag `Raw()` can also be omitted, since `Raw` is the default file format. Instead of using the functions **read** and **write** to access raw data, one can also use the operators << and >>.

Many file formats like `Fasta` or `Embl` are designed to store multiple records. For loading all records, one can call the function **read** repeatedly:

```
fstream fstrm;
fstrm.open("ests.fa", ios_base::in | ios_base::binary);
String<Dna> est;
while (! strm.eof())
{
    read(fstrm, est, Fasta()); //use sequence data in est
}
```

The function **goNext** skips the current record and proceeds to the next record. Each record contains a piece of data (i.e., a sequence or an alignment) and some optional, additional metadata. One can load these metadata before (not after) loading the actual data using **readMeta**:

```
FILE * cstrm = fopen("genomic_data.embl", "r");

goNext(cstrm, Embl());
//skip first data record

String<Dna> dna_sequence;
read(cstrm, dna_sequence, Embl());   //reads second record

String<char> meta_data;
readMeta(cstrm, meta_data, Embl()); //reads third meta data
read(cstrm, dna_sequence, Embl());   //reads third record

fclose(cstrm);
```

The function **write** is used to write a record into a file. Depending on the file format, a suitable metadata string must also be passed to **write**. The following example program:

```
FILE * cstrm = fopen("genomic_data.fa", "w");
write(cstrm, "acgt", "the metadata", Fasta());
fclose(cstrm);
```

creates the following file `"genomic_data.fa"`:

```
>the metadata
ACGT
```

7.11.2 File Reader Strings

The most simple way of reading a file that contains sequence data is to use a *file reader string* that emulates a constant string on a given file. It is implemented in the specialization **FileReader** of **String**:

```
String<Dna, FileReader<Fasta> > fr("ests.fa");
//opens "ests.fa" for reading
cout << length(fr);
//prints length of the sequence
```

File reader strings support almost the complete string interface, including iteration. A file reader string should nevertheless be read sequentially, because random accesses can be very time consuming. Note that the contents of a file reader string cannot be changed. The constructor of the file reader string can also take a file from which the sequences will be loaded. For example, the following code will read the second sequence in the file:

```
FILE * cstrm = fopen("genomic_data.embl", "r");

goNext(cstrm, Embl());

String<char> meta_data;
readMeta(cstrm, meta_data, Embl());
//reads meta data of second record

String<Dna, FileReader<Embl> > fr(cstrm);
//reads sequence data of second record

fclose(cstrm);
```

Chapter 8

Alignments

An alignment (see Section 1.2.3) is a compact notation for the similarities and the differences between two or more sequences. To get the similar regions together, the alignment process allows the insertion of *gaps* into the sequences, so we will first discuss in Section 8.1 which data structures for storing *gapped sequences* are provided in SeqAn, before we propose the alignment data structures in Section 8.2. Algorithms for computing (global) alignments are explained in Sections 8.4 to 8.6.

8.1 Gaps Data Structures

In alignments obviously arises the need to store *gapped sequences*, i.e., sequences that contain gaps between the values. The simplest way to store a gapped sequence is to store it as a usual sequence using a value type that is extended by an extra blank value '-'. The gaps in the sequence are then represented by the regions in the sequences that contain blank values. On the other hand, it could be advantageous to store the *ungapped* sequence and the position of the gaps separately for at least three reasons: (1) this way, we can extend arbitrary sequence data structures to gapped sequences without copying the sequence, (2) one sequence – or parts of it – can participate in several alignments, and (3) storing gaps as sequences of blank values could be very expensive for long gaps or for repeated manipulation of the gaps.

The class `Gaps` implements in SeqAn data structures for storing the positions of gaps or, more precisely spoken, *gap patterns*. A *gap pattern* of a sequence $a = a_1 a_2 \ldots a_n$ is a strictly monotonically

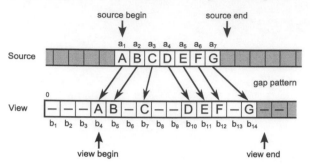

Figure 22: **Gaps Data Structure.** This is an example for a gap pattern function: $p(1) = 4$, $p(2) = 5$, $p(3) = 7$, ...

increasing function p that maps the values $\{1, 2, \ldots, n\}$ to values in $\mathbb{N}$ (see Figure 22). The sequence a is in general a segment of a larger host sequence s, which is called the *source* of p. The *source position* of a_i is its position within the source sequence s, e.g., if $a = s$, then the source position of a_i is $i - 1$.

For $p(i) = j$, we call $j - 1$ the *view position* of a_i. A j that is not a view position of any value in a is called a *blank*. A maximal run of blanks is called a *gap*. For example the L blanks $j+1, j+2, \ldots, j+ L$ between $p(i) = j$ and $p(i+1) = j+1+L$ are a gap of length L. Given a sequence a and a gap pattern p, the corresponding *gapped sequence* $b = b_1 b_2 \ldots b_m$ of length $m = p(n)$ is defined by:

$$
b_j = \begin{cases} a_i & \text{if there is an } i \text{ such that } p(i) = j \\ \text{`-'} & \text{(i.e., a blank character) otherwise} \end{cases}
$$

The gapped sequence is also called the *view* of the gap pattern. If there are blanks $0, 1, \ldots, p(1) - 1$ at the beginning of b, they are called the *leading gap* of b. We call $p(1) - 1$ the *begin view position* and $p(n)$ the *end view position* of p, i.e., they are the begin and end position of the gapped sequence without the leading gap. The gapped sequence b can also be thought to be followed by an endless number of trailing blanks $b_{m+1} = b_{m+2} = \ldots = \text{`-'}$. We call these trailing blanks the *trailing gap*. A run of values without blanks between two gaps is called a *non-gap*.

SeqAn offers three different specializations for `Gaps`, each of which has certain advantages: (1) `SequenceGaps`Gaps, (2) `ArrayGaps`Gaps, and (3) `SumlistGaps`Gaps (see Table 10). All

SequenceGaps	A sequence of values including blank signs '-'.
ArrayGaps	The lengths of gaps and non-gaps are stored in an array.
SumlistGaps	The gap pattern is stored in a two-dimensional sum list.

Table 10: **Gaps Specializations Overview.**

implementations of gaps data structures offer functions for inserting and deleting gaps, for changing a gap's size, and for converting view positions to source positions and vice versa, which is necessary for example for random access of the values in the gapped sequence given a view position. Table 11 lists some common funcions and metafunction applicable to gapped sequences. The differences between the three specializations will be described now.

8.1.1 SequenceGaps Specialization

The SequenceGaps specialization of Gaps is the most obvious implementation of a gaps data structure: It stores a gapped sequence simply as a sequence including '-' blank values. Each gap of size L takes therefore L blank values, except for the leading and the trailing gaps that are known because SequenceGaps explicitly saves the begin view position and the length of the sequence in member variables. This special treatment of leading gaps makes sense because the leading and trailing gaps could be very long, especially if the Gaps data structure is used to store a line in a multiple sequence alignment as described in Section 8.2. Since SequenceGaps stores the source together with the gaps, it cannot refer to an external source sequence. If the *ungapped* source is accessed by calling the function **source**, a temporary source string is created and returned by value.

The specialization SequenceGaps stores the view of the gapped sequence directly, so both iterating and random accessing its values are very fast. However, inserting and deleting blanks could be very expensive, because for each such operation the whole part of the view behind the modified position must be moved. A conversion between view and source position is also rather time

Metafunctions	
Source	Type of the underlying *ungapped* sequence, which can be accessed by the function **source** and set by **setSource** or **assignSource**.

Functions	
source	Returns the source of the gapped sequence, i.e., the underlying ungapped sequence.
assignSource	Assigns the source of the gapped sequence. The source sequence is copied into the gapped sequence.
setSource	Sets the source sequence; makes the gapped sequence *dependent* from this source sequence.
sourceBegin	The source begin position.
sourceEnd	The source end position.
insertGaps	Insert one or more blanks into the gapped sequence at a given position.
removeGaps	Removes blanks from the gapped sequence at a given position.
clearGaps	Removes all gaps from gapped sequence.
countGaps	Returns the number of gaps.
isGap	Determines whether there is a blank at given position.

Table 11: **Common Functions and Metafunctions for Gapped Sequences**. Gapped sequences also support the functions and metafunctions for ungapped sequences; see Table 4.

consuming, because it involves a linear scan through the gapped sequence. Although `SequenceGaps` is rather space efficient for gapped sequences that only contain short gaps, `SequenceGaps` becomes wasteful for very large gaps, because the space needed to store a gap is linear to its size.

8.1.2 ArrayGaps Specialization

The `ArrayGaps` specialization of `Gaps` stores the sizes of gaps and gap free parts. For example, for the gapped sequence in Figure 22, it stores the array $\{3, 2, 1, 1, 2, 3, 1, 1\}$. Every second number in this array corresponds to a non-gap's size, the rest corresponds to the lengths of the gaps. The first number in the array is the size of the leading gap. The source sequence can either be stored within the `Gaps` object or separately.

The advantage of `ArrayGaps` is to store the gaps more space efficient than `SequenceGaps` even for very large gaps.

8.1.3 SumlistGaps Specialization

The `SumlistGaps` specialization of `Gaps` stores a sequence of pairs, one pair for each non-gap in the gapped sequence. Each pair stores (1) the size of the non-gap and (2) the size of the non-gap plus the size of the preceding gap. For example, for the gapped sequence in Figure 22, it stores the following sequence of pairs: $\{(2, 5), (1, 2), (3, 5), (1, 2)\}$. These pairs are saved in a two-dimensional *sum list*. Given a sum S and a dimension d, the sum list $pair_1, pair_2, \ldots, pair_t$ allows a fast search for the first pair $pair_i$ in the list, for which holds: The d-th dimension of $sum_i := pair_1 + pair_2 + \ldots + pair_i$ is greater or equal to S. The search returns both $pair_i$ and sum_i. Note that sum_i is a pair of source position $sum_i[0]$ and view position $sum_i[1]$ for the first blank behind the i-th non-gap. The conversion between view position and source position works as follows:

(1) **source $\Longrightarrow$ view**:

Given a source position S, search the first dimension of the sum list for S and find $pair_i$. The view position V that corresponds to S is given by:

$$V = sum_i[1] - (sum_i[0] - S)$$

(2) **view** $\implies$ **source**:

Given a view position V, search the second dimension of the sum list for V and find $pair_j$. The source position S that corresponds to V is given by:

$$S = \begin{cases} sum_j[0] - pair_j[0] & \text{if } V \text{ is the position of a blank,} \\ sum_j[0] - (sum_j[1] - V) & \text{otherwise} \end{cases}$$

All operations on a sum list that are relevant for the implementation of `SumlistGaps` – like searching, inserting a pair, removing a pair, and changing a value of a pair – take $O(\log t)$ time, where t is the number of pairs in the list. This affects the runtime of operations on `SumlistGaps` gapped sequences: Inserting or removing gaps, changing a gap's size, conversion between view position and source position, and accessing the value at a given view position take time logarithmic to the number of gaps; see Table 12 and Figure 23.

8.2   Alignment Data Structures

Let us write a set of gap patterns $\{p_1, p_2, \ldots, p_k\}$ for $k \geq 2$ sequences $a^1, a^2, \ldots, a^k$ in a matrix, i.e., the rows are the gap patterns and the columns the view positions; see Figure 24.

A position j is called a *gap column*, if it is a blank in all gap patterns. Moreover, if j is a part of the leading gaps of all p_i, then j is a *leading gap column*, and if j is a part of the trailing gaps of all p_i, then j is a *trailing gap column*. The set $\{p_1, p_2, \ldots, p_k\}$ is called an *alignment* $\mathcal{A}$ of the sequences $a^1, a^2, \ldots, a^k$, if it contains no gaps columns but, potentially, leading and trailing gap columns. We say that values are *aligned*, if they belong to the same column. We can transform an arbitrary set of gap patterns into an

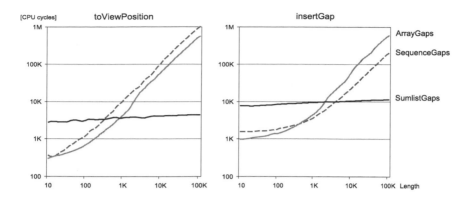

Figure 23: **Gaps Data Structure Run Times. Left:** Run times for converting source to view position of the *last* value depending on the total number of gaps in the gapped sequence. **Right:** Run times for inserting a gap at the *front* of a gapped sequence. In both cases, we used a gapped sequence of minimal length, which is *best-case* for SequenceGaps.

	SequenceGaps	ArrayGaps	SumlistGaps
Inserting a new gap	$O(n)$	$O(g)$	$O(\log(g))$
Removing a gap	$O(n)$	$O(g)$	$O(\log(g))$
Changing a gap's size	$O(n)$	$O(1)$	$O(\log(g))$
Conversion view to source	$O(n)$	$O(g)$	$O(\log(g))$
Conversion source to view	$O(n)$	$O(g)$	$O(\log(g))$
Accessing the value at a given view position	$O(1)$	$O(g)$	$O(\log(g))$
Accessing the value of an iterator	$O(1)$	$O(1)$	$O(1)$
Moving an iterator to the next position	$O(1)$	$O(1)$	$O(1)$
Moving an iterator to a given view position	$O(1)$	$O(g)$	$O(\log(g))$

Table 12: **Time Consumption of Operations on Gapped Sequences.** n is length of the gapped sequence, g is the number of gaps in the gapped sequence.

gap column

a^1 - AC - - AAG - CGTAGCA - -

a^2 - AC - TACGA - G - AGCA - -

a^3 CACTTATG - CC - AG - - - -

Figure 24: **Example of an Alignment.** An alignment of three sequences a^1, a^2, and a^3. (Note that the gap column has to be removed in order to get a proper alignment.)

alignment by removing all gap columns. For each proper subset $M \subset \{1, 2, \ldots, k\}$, a *projection* $\mathcal{A}_M$ of the alignment $\mathcal{A}$ is defined as the set $\{p_i \mid i \in M\}$ after removing all gap columns. There are two ways for storing alignments in SeqAn: (1) the `Align` data structures, and (2) alignment graphs (see Section 12.2).

Metafunctions	
Row	The type of an alignment row, typically a gapped sequence (Section 8.1).
Col	The type of an alignment column.
Functions	
rows	Returns the sequence of alignment rows.
cols	Returns the sequence of alignment columns.
setStrings	Uses the strings in a `StringSet` as rows of the alignment.
globalAlignment	Applies a global alignment algorithm (Section 8.5).
globalAlignment	Applies a local alignment algorithm (Section 10.1).

Table 13: **Common Functions and Metafunctions for Alignments**.

The data structure `Align` is implemented as a sequence of `Gaps` objects that store the rows of the alignment and are accessible via the function `rows`. Some typical functions and metafunctions for `Align` are listed in Table 13. An alignment can also be considered as sequence of columns, which can be retrieved using the function `cols`. The smallest position that is not a gap column is called the begin position, and the position of the first trailing gap column is called the end position of the columns sequence. The iterator of the column sequence is implemented as a set of k iterators, one iterator for each row. This means that iterating the column sequence of an alignment could be costly for alignments that contain many sequences.

Note that `Align` objects support gap columns, so it is the user's responsibility to remove them if necessary.

8.3 Alignment Scoring

8.3.1 Scoring Schemes

A *scoring scheme* for alignments is a function that maps alignments to numerical scores like `int` or `double` values. SeqAn supports alignment scoring schemes that are defined (1) by a function α that scores pairs of aligned values and (2) a function γ for scoring gaps. A gap of size l scores:

$$\gamma = g_{\text{open}} + g_{\text{extend}} * (l - 1), \tag{8.1}$$

i.e., the first blank in the gap scores g_{open}, and g_{extend} is added to γ for each further blank in the gap. Usually, we demand $g_{\text{open}} \leq g_{\text{extend}} \leq 0$, so γ is a convex function and larger gaps get a discount. If $g_{\text{open}} = g_{\text{extend}}$, then we call γ *linear*, otherwise γ is *affine*.

`Simple`	Defines the function α by two values a_{match} and $a_{mismatch}$ as follows: $\alpha(x, y) = a_{match}$, if $x = y$, otherwise $\alpha(x, y) = a_{mismatch}$. If no other values are specified, this scoring scheme implements the *edit distance* (Equation 8.4).
`ScoreMatrix`	Stores the function α in a matrix, which can be loaded from a file. Some common scoring matrices for `AminoAcid` values by Henikoff and Henikoff (1992) are predefined: `Blosum30`, `Blosum62`, and `Blosum80`. The matrix can be loaded by the function `read` from a file and stored by the function `write` to a file; see Section 7.11.
`Pam`	This is series of common scoring schemes for `AminoAcid` values by Dayhoff, Schwartz, and Orcutt (1978). A variant by Jones, Taylor, and Thornton (1992) is also available.

Table 14: **Alignment Scoring Schemes.** Specializations for `Score`.

The class `Score` implements some scoring schemes; see Table 14. If $\mathcal{A}$ is an alignment of two sequences a^1 and a^2, then we define the score of $\mathcal{A}$ by:

$$score(\mathcal{A}) := \sum_{\substack{\text{aligned values } (x,y) \text{ in } \mathcal{A}}} \alpha(x,y) + \sum_{\substack{\text{gaps } g \text{ in } \mathcal{A}}} \gamma(g) \qquad (8.2)$$

For alignments $\mathcal{A}$ of more than two sequences, we define the sum of pairs *score* to be the sum of the scores of all pairwise sub-alignments:

$$score(\mathcal{A}) := \sum_{i \neq j} score(\mathcal{A}_{\{i,j\}}) \qquad (8.3)$$

8.3.2 Sequence Similarity and Sequence Distance

Based on alignment scoring, we define a similarity measure for sequences as follows: The *sequence similarity* $sim(a^1, a^2)$ between two sequences a^1 and a^2 with respect to a given scoring scheme *score* is the maximum score alignments between a^1 and a^2 can get, i.e.,:

$$sim(a^1, a^2) := \max \left(score(\mathcal{A}) \text{ where } \mathcal{A} \text{ aligns } a^1 \text{ and } a^2 \right)$$

An alignment of a^1 and a^2 with score $sim(a^1, a^2)$ is called an *optimal alignment* of a^1 and a^2. We will describe some algorithms for computing optimal alignments in the next sections.

Note that all scoring schemes in SeqAn are meant to be "the higher, the better," that is alignment algorithms always try to *maximize* scores. However, we can also apply these algorithms for *minimizing* scores simply by maximizing their *negative values*: Let $\mathcal{A}^*$ be an alignment that scores *minimal* with respect to a scoring scheme *score*, then it is easy to prove that $\mathcal{A}^*$ also scores *maximal* with respect to the scoring scheme *score'* that is defined by: $score'(\mathcal{A}) := -score(\mathcal{A})$ for each alignment $\mathcal{A}$.

A *minimal* alignment score $score'(\mathcal{A}^*)$ can be seen as a *distance* between two sequences, so we define the *sequence distance* $dist(a^1, a^2)$ of two sequences a^1 and a^2 to be the negative value of their similarity:

$$dist(a^1, a^2) := -sim(a^1, a^2)$$

A well-known example of a sequence distance metric called *edit distance* or *Levenshtein distance* (Levenshtein 1965) is defined by the following scoring scheme:

$$\alpha(x, y) = \begin{cases} 0, \text{ if } x = y \\ -1, \text{ otherwise} \end{cases} \qquad g_{\text{open}} = g_{\text{extend}} = -1. \qquad (8.4)$$

8.4 Alignment Problems Overview

The *alignment problem* means to find an alignment with optimal score in the space of all possible alignments between two or more sequences. There are some variants of alignment problems:

(1) **Global Alignment Problem.** Alignments between complete sequences are *global alignments*. One way of solving the global alignment problem is *dynamic programming*, which is discussed in the next Section 8.5.

(2) **Maximum Weight Trace Problem.** Finding an optimal subgraph of a given alignment graph that is compatible with some optimal alignment (i.e., a *trace*) is called the *maximum weight trace problem*. We will discuss this in Section 12.2.2.

(3) **Local Alignment Problem.** A *local alignment* between two sequences a and b is a global alignment between a substring of a and substring of b, and the *local alignment problem* is to find an optimal local alignment. Local aligning is therefore a kind of *motif finding*; we will discuss it in Section 10.1.

(4) **Semi-Global Alignment.** A mix between global and local aligning is the so-called *semi-global alignment problem* that means globally aligning two sequences where some start or end gaps are free. One example for semi-global alignment is *overlap alignments*, that is finding the best possible alignment between a suffix of one sequence and a prefix of the other sequence. We will show in Section 8.5.4 how the user

can decide in SeqAn what start gap or end gap should be free when aligning two sequences.

8.5   Global Alignments

The *global alignment* problem is defined as follows: For a given set of sequences $a^1, a^2, \ldots, a^k$, find an alignment $\mathcal{A}^*$ of these sequences that scores optimal with respect to a given scoring scheme. Finding an optimal alignment of multiple sequences using sum-of-pair scoring (Section 8.3.1) is known to be NP-hard (Wang and Jiang 1994), but for a fixed number k of sequences, the alignment problem can be solved in time $O(n^k)$, where n is the length of the sequences.

NeedlemanWunsch	A dynamic programming algorithm by Needleman and Wunsch (1970) for linear gap costs. It aligns two sequences in quadratic time and space.
Gotoh	An extension of the Needleman-Wunsch algorithm that can deal with affine gap costs. (Gotoh 1982)
Hirschberg	A linear space dynamic programming algorithm. (Hirschberg 1975)
MyersHirschberg	A combination of the bit parallel algorithm by Myers (1999) and Hirschberg's algorithm. It aligns two sequences in linear space using edit distance.

Table 15: **Global Alignment Algorithms.** These algorithms are based on *dynamic programming.*

Algorithms for finding good global alignments in SeqAn can be accessed by calling the `globalAlignment` function. This function has as arguments (1) an `Align` object, alignment graph object or stream that will be used to store or display the found alignment, (2) a string set that contains the strings to align, if the strings are not already defined by the first argument, (3) a scoring scheme, and (4) a tag that specifies the algorithms that will be used for aligning; see Table 15. The function returns the score of the computed alignment.

	linear gap costs		affine gap costs	
	time (s)	space (MB)	time (s)	space (MB)
SeqAn				
Needleman-Wunsch	3.3	236	6.3	236
Hirschberg			14.7	4
Myers-Hirschberg	**0.2**	3		
NCBI C++ toolkit				
Needleman-Wunsch			**4.0**	245
Hirschberg			6.6	14
Bio++	**13.4**	2100	28.0	≈6000
BTL			**96162**	933
BioJava	**76**	2000	93	≈6000

Table 16: **Runtimes and Internal Space Requirements for Computing Sequence Alignments.** The table shows average time and space requirements for aligning the genomes of two human influenza viruses, each of length about 15.6 *kbp*, using alignment functions in SeqAn, the NCBI C++ toolkit (Vakatov et al. 2003), Bio++ (Dutheil et al. 2006), BTL (Pitt et al. 2001), and BioJava (Holland et al. 2008). Runtimes printed in boldface show for each library the time of the fastest algorithm for computing an alignment using *edit distance*.

8.5.1 Needleman-Wunsch Algorithm

In 1970, Needleman and Wunsch introduced an algorithm based on *dynamic programming* (Bellman 1957) to solve the global alignment problem with linear gap costs for two sequences $a = a_1 \ldots a_n$ and $b = b_1 \ldots b_m$. This algorithm is based on the following observation: Let $\mathcal{A}_{i,j}$ be an optimal alignment of the prefixes $a_1 \ldots a_i$ and $b_1 \ldots b_j$ for $i \in \{1, \ldots, n\}$ and $j \in \{1, \ldots, m\}$, and $M_{i,j} = score(\mathcal{A}_{i,j})$. Then the alignment $\mathcal{A}'_{i,j}$ that we get after deleting the last column $\mathcal{C}$ from $\mathcal{A}_{i,j}$ is an optimal alignment, and

$$score(\mathcal{A}_{i,j}) = score(\mathcal{A}'_{i,j}) + score(\mathcal{C}). \tag{8.5}$$

There are three cases: (1) $\mathcal{C}$ aligns a_i and b_j, then $score(\mathcal{A}'_{i,j}) = M_{i-1,j-1}$, or (2) $\mathcal{C}$ aligns a_i to a blank, then $score(\mathcal{A}'_{i,j}) = M_{i-1,j}$, or (3) $\mathcal{C}$ aligns b_j to a blank, then $score(\mathcal{A}'_{i,j}) = M_{i,j-1}$. Therefore we can compute $M_{i,j}$ according to the recursion:

$$M_{i,j} \leftarrow \max \begin{cases} M_{i-1,j-1} + \alpha(a_i, b_j) \\ M_{i-1,j} + g \\ M_{i,j-1} + g \end{cases} \tag{8.6}$$

where $\alpha(a_i, b_j)$ is the score for aligning a_i and b_j, g is the score for a blank, and $M_{i,0} = i*p$, and $M_{0,j} = j*p$. Algorithm 2 enumerates all pairs (i, j) for $1 \leq i \leq n$ and $1 \leq j \leq m$ in increasing order for i and j, so $M_{i-1,j-1}$, $M_{i-1,j}$, and $M_{i,j-1}$ are already known before $M_{i,j}$ is computed. FILLMATRIX also protocols in $T_{i,j}$ which of the three cases was applied to compute $M_{i,j}$. This information is used in TRACEBACK to construct an optimal alignment. The overall time consumption is $O(n \times m)$, and since TRACEBACK requires a complete score matrix T, the space requirements are also $O(n \times m)$. An example of the dynamic programming matrix M is shown in Figure 25.

There is no need to store the complete matrix M during the execution of FILLMATRIX, because at any time at most $m+1$ cells of M are needed for proceeding: After computing $M_{i,j}$, only the values in $M_{i,j-1}, \ldots, M_{n,j-1}, M_{1,j}, \ldots M_{i,j}$ play a role for the rest of the computation. We can therefore adapt FILLMATRIX to compute the optimal score $M_{n,m}$ in linear space.

Note that it is possible to generalize the Needleman-Wunsch algorithm in a way that it can compute optimal alignments for arbitrary gap costs in time $O(n^3)$, but this algorithm is quite slow and hence rarely applied in practice, so it is not provided by SeqAn.

Figure 25: **Needleman-Wunsch Algorithm.** The dynamic programming matrix for aligning $a =$"AATCTAGCGT" and $b =$"GTACATTTGACG". The values of M for edit distance scoring, and T is visualized by pointers to the best predecessors. The optimal alignment on the bottom corresponds to the black printed path, its score -7 is the value in the lower right cell. The gray fields are part of back traces for alternative optimal alignments.

$\triangleright$ NEEDLEMANWUNSCH$(a_1 \ldots a_n, b_1 \ldots b_m)$
1    $(M, T) \leftarrow$ FILLMATRIX$(a_1 \ldots a_n, b_1 \ldots b_m)$
2   **return** TRACEBACK$(a_1 \ldots a_n, b_1 \ldots b_m, T)$

$\triangleright$ FILLMATRIX$(a_1 \ldots a_n, b_1 \ldots b_m)$
1    $M_{0,0} \leftarrow 0$
2    $M_{i,0} \leftarrow i * y$ for $i \in \{1, \ldots, n\}$ $\left.\begin{array}{l} \\ \\ \end{array}\right\}$ initiali-
3    $M_{0,j} \leftarrow j * g$ for $j \in \{1, \ldots, m\}$ zation
4   **for** $i \leftarrow 1$ **to** n **do**
5   **for** $j \leftarrow 1$ **to** m **do**

$$6 \qquad M_{i,j} \leftarrow \max \begin{cases} M_{i-1,j-1} + \alpha(a_i, b_j) & = case_{diag} \\ M_{i-1,j} + g & = case_{up} \\ M_{i,j-1} + g & = case_{left} \end{cases}$$

7      $T_{i,j} \leftarrow \operatorname{argmax}_k case_k$
8   **return** (M, T)

$\triangleright$ TRACEBACK$(a_1 \ldots a_i, b_1 \ldots b_j, T)$
1   **case** $i = j = 0$: $\left.\begin{array}{l} \\ \\ \\ \end{array}\right\}$ break
2   **return** $\begin{bmatrix} \ \end{bmatrix}$ condi-
     tion

3   **case** $T_{i,j} = up$ or $j = 0$:
4   **return** $\left[\text{TRACEBACK}(a_1 \ldots a_{i-1}, b_1 \ldots b_j, T) \, \middle| \, \begin{matrix} a_i \\ - \end{matrix} \right]$

5   **case** $T_{i,j} = left$ or $i = 0$:
6   **return** $\left[\text{TRACEBACK}(a_1 \ldots a_i, b_1 \ldots b_{j-1}, T) \, \middle| \, \begin{matrix} - \\ b_j \end{matrix} \right]$ recursion

7   **case** $T_{i,j} = diag$:
8   **return** $\left[\text{TRACEBACK}(a_1 \ldots a_{i-1}, b_1 \ldots b_{j-1}, T) \, \middle| \, \begin{matrix} a_i \\ b_j \end{matrix} \right]$

Algorithm 2: **Needleman-Wunsch Algorithm.** $\alpha(a_i, b_j)$ is the score for aligning a_i and b_j, g is the score for a blank.

8.5.2 Gotoh's Algorithm

The algorithm by Needleman and Wunsch (Section 8.5.1) does not work for *affine gap costs*, i.e., if gaps of length l score $g_{open} + g_{extend} * (l - 1)$ with $g_{open} \neq g_{extend}$. Let C be the last column of an alignment $\mathcal{A}_{i,j} = \mathcal{A}'_{i,j}C$. If C extends a gap of $\mathcal{A}'_{i,j}$, then $score(\mathcal{A}_{i,j}) = score(\mathcal{A}'_{i,j}) + g_{extend} \neq score(\mathcal{A}'_{i,j}) + g_{open}$, hence Equation 8.5 does not hold anymore.

To deal with affine gap costs, Gotoh adapted the algorithm in 1982 such that it computes for each $i \in \{1, \ldots, n\}$ and $j \in \{1, \ldots, m\}$ the following three scores of alignments between $a_1 \ldots a_i$ and $b_1 \ldots b_m$: (1) the optimal alignment score $M_{i,j}$, (2) the best score $I^a_{i,j}$ of alignments that align a_i to a blank, and (3) the best score $I^b_{i,j}$ of alignments that align b_j to a blank. This can be done by modifying FILLMATRIX as it is shown in Algorithm 3. The asymptotic time and space requirements are the same as for the algorithm by Needleman and Wunsch ($O(n \times m)$) but with larger constant factors, since the algorithm by Gotoh must store and fill three matrices instead of one.

If $\alpha(a_i, b_j) \geq g_{open} + g_{extend}$ for any a_i and b_j, then the algorithm can also be implemented by only two matrices M and I, where $I_{i,j}$ stores the best score of alignments $a_1 \ldots a_i$ and $b_1 \ldots b_m$ that either align a_i or b_j to a blank. This optimization is currently not provided by SeqAn.

8.5.3 Hirschberg's Algorithm

Unlike the Needleman-Wunsch algorithm (Section 8.5.1) or Gotoh's algorithm (Section 8.5.2), which both take space $O(n \times m)$ to compute optimal alignments of two sequences $a = a_1 \ldots a_n$ and $b = b_1 \ldots b_m$, the algorithm by Hirschberg (1975) only needs linear space. Hirschberg's algorithm (Algorithm 4) applies a divide-and-conquer strategy: It splits both a and b into two parts and aligns them separately.

The algorithm is illustrated in Figure 26. The sequence a is cut at position $i = \lfloor \frac{n}{2} \rfloor$ (for $n > 1$) into two halves $a_1 \ldots a_i$ and $a_{i+1} \ldots a_n$. The main problem is to find an appropriate cutting position j in b, such that an optimal alignment between a and b exists that

$\triangleright$ FILLMATRIXGOTOH$(a_1 \ldots a_n, b_1 \ldots b_m)$

1. $M_{0,0} \leftarrow 0$
2. $M_{i,0} \leftarrow i * g_{\text{extend}}$ for $i \in \{1, \ldots, n\}$
3. $M_{0,j} \leftarrow j * g_{\text{extend}}$ for $j \in \{1, \ldots, m\}$
4. $I^a_{0,j} \leftarrow -\infty$ for $j \in \{1, \ldots, m\}$
5. $I^b_{i,0} \leftarrow -\infty$ for $i \in \{1, \ldots, n\}$
6. **for** $i \leftarrow 1$ **to** n **do**
7.   **for** $j \leftarrow 1$ **to** m **do**
8. $I^a_{i,j} \leftarrow \max \begin{cases} M_{i-1,j} + g_{\text{open}} \\ I^a_{i-1,j} + g_{\text{extend}} \end{cases}$
9. $I^b_{i,j} \leftarrow \max \begin{cases} M_{i,j-1} + g_{\text{open}} \\ I^b_{i,j-1} + g_{\text{extend}} \end{cases}$
10. $M_{i,j} \leftarrow \max \begin{cases} M_{i-1,j-1} + \alpha(a_i, b_j) & = case_{diag} \\ I^a_{i,j} & = case_{up} \\ I^b_{i,j} & = case_{left} \end{cases}$
11. $T_{i,j} \leftarrow \text{argmax}_k \; case_k$
12. **return** (M, T)

(braces spanning lines 1–5: initialization)

Algorithm 3: **The Recursion of Gotoh's Algorithm.**

$\triangleright$ HIRSCHBERG$(a_1 \ldots a_n, b_1 \ldots b_m)$

1. **if** $n < 2$ **then**
2. $\mathcal{A} \leftarrow$ NEEDLEMANWUNSCH$(a_1 \ldots a_n, b_1 \ldots b_m)$
3. **else**
4. $i \leftarrow \lfloor \frac{n}{2} \rfloor$
5. $M^{\mathcal{L}} \leftarrow$ FILLMATRIX$(a_1 \ldots a_i, b_1 \ldots b_m)$
6. $M^{\mathcal{R}} \leftarrow$ FILLMATRIX$(a_n \ldots a_{i+1}, b_m \ldots b_1)$
7. $t \leftarrow \text{argmax}_j (M^{\mathcal{L}}_{i,j} + M^{\mathcal{R}}_{n-i,m-j})$
8. $\mathcal{L} \leftarrow$ HIRSCHBERG$(a_1 \ldots a_i, b_1 \ldots b_t)$
9. $\mathcal{R} \leftarrow$ HIRSCHBERG$(a_{i+1} \ldots a_n, b_{t+1} \ldots b_m)$
10. $\mathcal{A} \leftarrow \mathcal{L}\mathcal{R}$
11. **return** $\mathcal{A}$

(braces: lines 5–7 find t; lines 8–9 recursion)

Algorithm 4: **Hirschberg's Algorithm.**

aligns $a_1 \ldots a_i$ to $b_1 \ldots b_j$ and $a_{i+1} \ldots a_n$ to $b_{j+1} \ldots b_m$. For any $j \in \{0, \ldots, m\}$, let $\mathcal{L}^j$ be an optimal alignment of the prefixes $a_1 \ldots a_i$ and $b_1 \ldots b_j$, and $\mathcal{R}^j$ an optimal alignment of the suffixes $a_{i+1} \ldots a_n$ and $b_{j+1} \ldots b_m$. There is a $t \in \{0, \ldots, m\}$ for which the combination $\mathcal{A}^t := \mathcal{L}^t \mathcal{R}^t$ is an optimal alignment of a and b. For finding a t that maximizes the total score $score(\mathcal{L}^t) + score(\mathcal{R}^t)$, we have to compute the scores of all $\mathcal{L}^j$ and $\mathcal{R}^j$. A single call

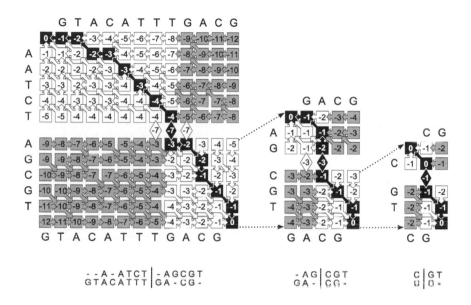

Figure 26: **Hirschberg's Algorithm.** This figure illustrates three recursion steps when aligning $a =$"AATCTAGCGT" and $b =$"GTACATTTGACG". The black printed path corresponds to the alignment that is about to be computed. The gray cells need not be recomputed during the next recursion step.

of FILLMATRIX$(a_1 \ldots a_i, b)$ in line 5 of HIRSCHBERG computes $M_{i,j}^{\mathcal{L}} = score(\mathcal{L}^j)$ for all j. The scores of the $\mathcal{R}^j$ are computed similarly in line 6 by passing the *reverses* of $a_{i+1} \ldots a_n$ and b to FILLMATRIX: The entry $M_{n-i,m-j}^{\mathcal{R}}$ of the computed matrix $M^{\mathcal{R}}$ is the optimal score for aligning $a_n \ldots a_{i+1}$ and $b_m \ldots b_{j+1}$, which is the same as the best score for aligning $a_{i+1} \ldots a_n$ and $b_{j+1} \ldots b_m$. FILLMATRIX only takes linear space for computing the scores needed. Hence, the total space requirement of HIRSCHBERG is $O(n + m)$. It is easy to prove by induction that HIRSCHBERG takes time $O(n \times m)$.

The implementation of Hirschberg's algorithm in SeqAn combines it with Gotoh's algorithm (Section 8.5.2) for sequence alignments using affine gap cost schemes.

The alignment algorithm `MyersHirschberg`, which is the fastest algorithm in SeqAn for global sequence alignment (see Table 16), can be used when aligning two sequences using *edit distance* scoring. This variant of Hirschberg's algorithm uses Myers' bitvector algorithm (see Section 9.3.2) instead of FILLMATRIX for computing the scores for $\mathcal{L}^j$ and $\mathcal{R}^j$.

8.5.4 Aligning with Free Start or End Gaps

After some simple modifications, both the Needleman-Wunsch algorithm and Gotoh's algorithm can also be used to compute alignments with free *start gaps* or *end gaps*. A *start gap* contains a blank that is aligned to a_1 or b_1, and an *end gap* contains a blank that is aligned to a_n or b_m. Gap scores are usually non-positive values, and we call a gap *free*, if it scores 0.

Start gaps in a become free when $M_{i,0}$ are set to 0 for all $i \in \{1, \ldots, n\}$ (FILLMATRIX, line 2). For free start gaps in b, we set $M_{0,j} = 0$ for $j \in \{1, \ldots, m\}$ (FILLMATRIX, line 3).

Let $i_{\max} = \text{argmax}_{i \in \{1,\ldots,n\}} M_{i,m}$ and $j_{\max} = \text{argmax}_{j \in \{1,\ldots,m\}} M_{n,j}$. The algorithm TRACEBACK$(a_1 \ldots a_{i_{\max}}, b)$ computes an optimal alignment $\mathcal{A}_{i_{\max}}$ of $a_1 \ldots a_{i_{\max}}$ and b. If end gaps in a – but not in b – are free, then this is also the best alignment of a and b. For free end gaps in b – but not in a –, the function call TRACEBACK$(a, b_1 \ldots b_{j_{\max}})$ returns the optimal alignment $\mathcal{A}_{j_{\max}}$ of a and b. If end gaps are free both in a and b, then either $\mathcal{A}_{i_{\max}}$ or $\mathcal{A}_{j_{\max}}$ is optimal, whichever is better.

The class `AlignConfig` can be used to specify, which start gap or end gap are free when calling `globalAlignment`. `AlignConfig` has four `bool` template arguments; `true` means *gap is free*. Listing 19 shows an example for using `AlignConfig`.

```
StringSet<CharString> string_set;
appendValue(string_set, a);
appendValue(string_set, b);
Align<CharString> alignment(string_set);
globalAlignment(alignment,
                Score<int>(),
                AlignConfig<false, false, true, false>(),
                NeedlemanWunsch());
```

Listing 19: **Example for Using** `AlignConfig`. The two sequences a and b
are aligned, end gaps for b are free.

8.5.5 Progressive Alignment

SeqAn also offers a progressive heuristic for finding good alignments between more than two sequences (Rausch, Emde, Weese,
Döring, Notredame, and Reinert 2008). We already described the
idea of this algorithm in Section 1.2.3 when we discussed the software tool CLUSTAL W (Thompson et al. 1994): The sequences
$a^1, \ldots, a^d$ are aligned step by step following a binary *guide tree* $\mathcal{T}$,
which is constructed by a *hierarchical clustering* algorithm (line 2
of PROGRESSIVEALIGN; see Algorithm 5), on the basis of the pairwise distances between the sequences. SeqAn supports *agglomer-
ative clustering* (complete linkage, single linkage and UPGMA; see
e.g., Sneath and Sokal 1973) and *neighbor-joining* (Saitou and Nei
1987). FOLLOWGUIDETREE aligns the sequences following the
guide tree from the leaves to the root. At each vertex v, the alignments $\mathcal{A}^l$ and $\mathcal{A}^r$ from both children of v are aligned (line 8),
where we conceive $\mathcal{A}^l$ and $\mathcal{A}^r$ as *sequences of columns*, so they can
be aligned by any pairwise global sequence alignment algorithm
like Needleman-Wunsch (see Section 8.5.1). The score $\alpha(c^l, c^r)$
for aligning two alignment columns c^l and c^r is defined as the
(weighted) sum of scores $\alpha(a^l, a^r)$, where $a^l \in c^l$ and $a^r \in c^r$ (*sum
of pairs score*). Inserting a gap into $\mathcal{A}^l$ or $\mathcal{A}^r$ means to insert a
gap column, i.e., inserting a gap into all sequences of $\mathcal{A}^l$ or $\mathcal{A}^r$,
respectively.

The progressive alignment idea was also used in the software tool

```
        ▷ PROGRESSIVEALIGN(a¹, ..., aᵈ)
 1      D[i, j] ← dist(aⁱ, aʲ) for all i, j ∈ {1, ..., d}
 2      T ← CLUSTERING(D)
 3      A ← FOLLOWGUIDETREE(a¹, ..., aᵈ, T)
 4      return A
```

```
        ▷ FOLLOWGUIDETREE(a¹, ..., aᵈ, T)
 1      v ← root of T
 2      if v is leaf then
 3      |  A ← the sequence aⁱ on v
 4      else
 5      |  Tˡ, Tʳ ← left and right subtrees below v
 6      |  Aˡ ← FOLLOWGUIDETREE(a¹, ..., aᵈ, Tˡ)
 7      |  Aʳ ← FOLLOWGUIDETREE(a¹, ..., aᵈ, Tʳ)
 8      |  A ← align Aˡ and Aʳ
 9      return A
```

Algorithm 5: **Progressive Alignment.** We omit the function CLUSTERING
    that applies a clustering algorithm to compute the guide tree T for a given
    distance matrix D.

T-Coffee by Notredame et al. (2000) that uses a much more elaborated scoring function α, which is computed from a given set
C of pairwise local or global alignments between the sequences
$a^1, \ldots, a^d$. T-Coffee first defines for any pair of characters b and
b' that stem from different sequences a^i and a^j an *individual* score
$\alpha(b, b')$ that depends on the total score of alignments $A \in C$ in
which b and b' are aligned. In a second step, T-Coffee uses a
method called *triplet extension*, that applies the following rule: If
two values b and b' are aligned in $A \in C$, and b' and b'' are aligned
in another alignment $A' \in C$, then we reinforce the score $\alpha(b, b'')$
that we get for aligning b and b''. The triplet extension helps to
find an agreement between the pairwise alignments $\in C$, and this
results in much better multiple alignments.

The implementation of T-Coffee in SeqAn stores these values α as
weights on the edges of an alignment graph (Section 12.2) between
$a^1, \ldots, a^d$, and the alignment problem can then be defined as a
maximum weight trace problem (Section 12.2.2).

8.6 Chaining

We saw in Section 8.5 that computing the best alignment between two sequences using dynamic programming takes quadratic time, and the alignment of $d > 2$ sequences even takes exponential time in d. Fortunately there are faster heuristics for finding good – but not necessarily optimal – alignments. One way is to search for highly similar substrings, so-called *seeds*, and to combine them in a process called *chaining*. A chain of seeds then can be used as a backbone for a *banded alignment* of the two sequences; see Section 8.6.4. We demonstrate this principle in all details for the algorithm LAGAN in Chapter 13.

This section concerns chaining seeds to *global* alignments. Similar techniques could also be used to get good *local* alignments, as we will see in Section 10.2.2. How to find seeds will be discussed in Chapter 10.

8.6.1 Seeds

Basically, a seed $\mathcal{S}$ is a set of non-empty segments $s^1, \ldots, s^d$ of sequences $a^1, \ldots, a^d$, where $d \geq 2$ is called the *dimension* of $\mathcal{S}$. We call $left_i(\mathcal{S})$ the begin position of s^i and $right_i(\mathcal{S})$ the end position of the segment s_i for $i \in \{1, \ldots, d\}$. According to the conventions stated in Section 7.1, the position of the value a_i in a sequence $a_1 \ldots a_m$ is $i - 1$, and the begin position of a segment $a_{left} \ldots a_{right}$ is $left - 1$ and the end position $right$.

SeqAn offers a class `Seed` for storing seeds; the specializations of this class are listed in Table 17. All seed types implement the functions `leftPosition` and `rightPosition` to access the begin and end positions of their segments, and the functions `setLeftPosition` and `setRightPosition` to set them. Moreover, each seed $\mathcal{S}$ stores the score $weight(\mathcal{S})$ of an optimal alignment between its segments, which can be retrieved by the function `weight` and set by the function `setWeight`. Chaining only requires information about the dimension, borders, and scores of the seeds, i.e., we need not to know the complete alignments.

`SimpleSeed`	A seed of dimension $d = 2$. This is the preferred seed type for seed merging, extending, and local chaining algorithms; see Section 10.2. The default specialization of `Seed`.
`MultiSeed`	A seed type of arbitrary dimension $d \geq 2$ that was designed for global chaining.

Table 17: **Specializations of class `Seed`.**

8.6.2 Generic Chaining

In the following let d be a fixed seed dimension. We say that $\mathcal{S}_j$ can be *appended* to another seed $\mathcal{S}_k$, if $\mathcal{S}_j$ is *right of* $\mathcal{S}_k$, that is if $right_i(\mathcal{S}_k) \leq left_i(\mathcal{S}_j)$ for all $i \in \{1, \ldots, d\}$. Given a set of seeds $\{\mathcal{S}_1, \ldots, \mathcal{S}_n\}$, we define the *top seed* $\mathcal{S}_0$ to be the seed with $left_i(\mathcal{S}_0) = right_i(\mathcal{S}_0) = 0$ for all $i \in \{1, \ldots, d\}$ and $weight(\mathcal{S}_0) = 0$. The *bottom seed* $\mathcal{S}_{n+1}$ is defined by $left_i(\mathcal{S}_{n+1}) = right_i(\mathcal{S}_{n+1}) = \max_j\{right_i(\mathcal{S}_j)\}$ and $weight(\mathcal{S}_{n+1}) = 0$. Note that all seeds can be appended to $\mathcal{S}_0$ and that $\mathcal{S}_{n+1}$ can be appended to all seeds. An ordered set $\mathcal{C} = \mathcal{S}_{j_1}, \mathcal{S}_{j_2}, \ldots, \mathcal{S}_{j_k}$ of seeds is called a *chain*, if $\mathcal{S}_{j_{i+1}}$ can be appended to $\mathcal{S}_{j_i}$ for each $i \in \{1, \ldots, k-1\}$. The *score* of a chain is defined by:

$$score(\mathcal{C}) = \sum_{i=1}^{k} weight(\mathcal{S}_{j_i}) + \sum_{i=1}^{k-1} gapscore(\mathcal{S}_{j_i}, \mathcal{S}_{j_{i+1}}),$$

where $gapscore(\mathcal{S}_{j_i}, \mathcal{S}_{j_{i+1}})$ is the (usually non-positive) score for appending $\mathcal{S}_{j_{i+1}}$ to $\mathcal{S}_{j_i}$.

The *global chaining problem* is to find a maximal scoring chain $\mathcal{C}$ that starts with $\mathcal{S}_0$ and ends with $\mathcal{S}_{n+1}$. GENERICCHAINING (Algorithm 6) solves this problem in time $O(dn^2)$ by *dynamic programming*. The algorithm computes for each seed $\mathcal{S}_j$ the *predecessor* $\mathcal{S}_k$ for which the chain $\mathcal{S}_0, \ldots, \mathcal{S}_k, \mathcal{S}_j$ gets the optimal score. The score of this chain is stored in M_j and the index k of the predecessor is stored in T_j. The best *global* chain is reconstructed in CHAINTRACEBACK by following T starting from $\mathcal{S}_{n+1}$. The algorithm applies a *sweep line* technique (Shamos and Hoey 1976) by sorting the seeds in line 1 of Algorithm 6. This guarantees that

```
     ▷ GENERICCHAINING(S₁,...,Sₙ)
  1  sort S₁,...,Sₙ in increasing order of right₁(Sᵢ)
  2  compute top seed S₀ and bottom seed Sₙ₊₁
  3  for j ← 1 to n + 1 do
  4      Mⱼ  ← gapscore(S₀, Sⱼ) + weight(Sⱼ)
  5      Tⱼ  ← 0
  6      for k ← 1 to j − 1 do
  7          if rightᵢ(Sₖ) ≤ leftᵢ(Sⱼ) for all i ∈ {2,...,d}
             then
  8              score ← Mₖ + gapscore(Sₖ, Sⱼ) + weight(Sⱼ)
  9              if score > Mⱼ then
 10                  Mⱼ  ← score
 11                  Tⱼ  ← k
 12  return CHAINTRACEBACK(S₀,...,Sₙ₊₁, n + 1, T)
```

compute
best
predecessor
for $\mathcal{S}_j$

```
     ▷ CHAINTRACEBACK(S₀,...,Sₙ₊₁, j, T)
  1  if j − 0 then
  2      return S₀
  3  else
  4      return CHAINTRACEBACK(S₀,...,Sₙ₊₁, Tⱼ, T), Sⱼ
```

Algorithm 6: **Generic Chaining Algorithm.** $\mathcal{S}_1,\ldots,\mathcal{S}_n$ is a set of d-dimensional seeds, $d \geq 2$. The algorithm computes a maximal global chain; its score is stored in M_{n+1}.

the seed S_j can only be appended to seeds S_k with $k < j$, hence M_k was already computed before it is used in line 8 to compute M_j.

The function `globalChaining` implements chaining algorithms in SeqAn. The actual algorithm is specified by a tag; see Listing 20 for an example.

```
String< Seed<int, MultiSeed> > seeds;
...
String< Seed<int, MultiSeed> > chain;
Score<int, Manhattan> scoring;
int score = globalChaining(seeds, chain, scoring);
```

Listing 20: **Global Chaining Example.** We omit the process of filling the container **seeds** with seeds.

8.6.3   Chaining Using Sparse Dynamic Programming

The algorithm GENERICCHAINING takes quadratic time, because it has to examine a linear number of predecessor candidates S_k for each seed S_j. We can improve this for some *gapscore* functions by using efficient data structures that allow to determine an optimal predecessor seed in sublinear time. This technique called *sparse dynamic programming* (Eppstein et al. 1992) may speed up chaining as long as the dimension d of the seeds is small compared to the number n of seeds. Figure 27 shows the gap scoring functions for which SeqAn implements optimized chaining algorithms. For example, Algorithm 7 (described in Gusfield 1997, pages 325–329) solves the chaining problem for $d = 2$ and *gapscore* $\equiv 0$ (i.e., the scoring scheme *Zero*) in time $O(n \log n)$.

SPARSECHAINING enumerates all positions $left_1(S_j)$ and $right_1(S_j)$ in increasing order. If the begin position of a seed S_j is processed (lines 9 and 10), then the optimal score M_j of chains ending in S_j is computed, and the algorithm appends S_j to a seed $S_k \in D$, where D is a set of potentially optimal

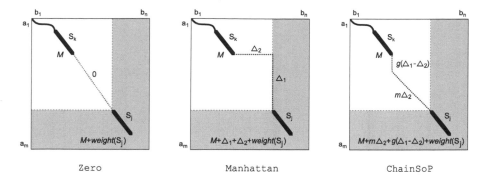

Zero   Manhattan   ChainSoP

Zero	All gaps between seeds score 0, that is $gapscore(\mathcal{S}_k, \mathcal{S}_j) = 0$ for all seeds $\mathcal{S}_k$ and $\mathcal{S}_j$.
Manhattan	The gap score is proportional to the sum of the distances between the segments in the seed, that is

$$gapscore(\mathcal{S}_k, \mathcal{S}_j) = g \sum_{i=1}^{d} \Delta_i,$$

where $\Delta_i = left_i(\mathcal{S}_j) - right_i(\mathcal{S}_k)$ and $g < 0$ is the score for a single blank.

ChainSoP   This gap scoring scheme was proposed by Myers and Miller (1995). For $d = 2$, the segments between $\mathcal{S}_k$ and $\mathcal{S}_j$ are aligned as long as possible with mismatches, and the rest is filled up with blanks, that is:

$$gapscore(\mathcal{S}_k, \mathcal{S}_j) = S_{1,2} = \begin{cases} m\Delta_2 + g(\Delta_1 - \Delta_2), & \text{if } \Delta_1 \geq \Delta_2 \\ m\Delta_1 + g(\Delta_2 - \Delta_1), & \text{if } \Delta_1 \leq \Delta_2 \end{cases}$$

where $\Delta_i = left_i(\mathcal{S}_j) - right_i(\mathcal{S}_k)$ and $g, m \leq 0$ are the scores for a single blank and a single mismatch.

For $d > 2$, *gapscore* is the sum-of-pairs score:

$$gapscore(\mathcal{S}_k, \mathcal{S}_j) = \sum_{1 \leq i < i' \leq d} S_{i,i'}$$

Figure 27: **Gap Scoring Schemes for Chaining.** The score of a gap between a seed S_j and a predecessor S_k; three specialization of class `Score` are listed. Note that $gapscore(\mathcal{S}_k, \mathcal{S}_j)$ is only defined if $\mathcal{S}_j$ can be appended to $\mathcal{S}_k$, i.e., if $\Delta_i = left_i(\mathcal{S}_j) - right_i(\mathcal{S}_k) \geq 0$ for all $i \in \{1, \ldots, d\}$.

predecessors for subsequent seeds. D is updated whenever the end position of a seed is processed (lines 12 to 15). Let $\mathcal{S}_j$ and $\mathcal{S}_k$ be two different seeds with $right_2(\mathcal{S}_j) \leq right_2(\mathcal{S}_k)$ and $M_j \geq M_k$. Then all subsequent seeds $\mathcal{S}_i$ that can be appended to $\mathcal{S}_k$ can also be appended to $\mathcal{S}_j$ without losing score. If $\mathcal{S}_j \in D$, then there is no need to keep $\mathcal{S}_k \in D$. We say that $\mathcal{S}_j$ *dominates* $\mathcal{S}_k$. A seed $\mathcal{S}_j$ is added to D in line 15, if and only if it is not dominated by any other seed in D, and in this case, all seeds $\mathcal{S}_k$ that are dominated by $\mathcal{S}_j$ are extracted from D in line 14. If follows that (1) D contains all seeds that were already processed except some seeds that are dominated by other seeds in D, and (2) no seed in D is dominated by another seed in D. Therefore D contains an optimal predecessor for any seed $\mathcal{S}_j$ that is about to be appended, and this is the seed $\mathcal{S}_{T_j} \in D$ found in line 9 because a *better* seed $\mathcal{S}_k \in D$ with $right_2(\mathcal{S}_k) \leq left_2(\mathcal{S}_j)$ and $M_k > M_{T_j}$ would dominate $\mathcal{S}_{T_j}$. Note that the results of the *argmax* operation in line 9 is well defined, because there is always a seed $\mathcal{S}_k \in D$ such that $right_2(\mathcal{S}_k) = 0$.

If we apply a suitable *dictionary* data structures for storing D like a *skip list* that sorts the seeds $\mathcal{S}$ according to $right_2(\mathcal{S})$, then each operation for searching, adding, and extracting seeds in D takes time $O(\log n)$. The complete algorithm runs in $O(n \log n)$, since each seed $\mathcal{S}$ is added to D and extracted from D only once.

We can apply SPARSECHAINING for the gap scoring scheme `Manhattan`, if we modify the condition for $\mathcal{S}_j$ *dominates* $\mathcal{S}_k$: Suppose that a seed $\mathcal{S}_l$ can be appended either to $\mathcal{S}_j$ and $\mathcal{S}_k$. Then $\mathcal{S}_j$ would be preferred, if $M_j + gapscore(\mathcal{S}_j, \mathcal{S}_l) > M_k + gapscore(\mathcal{S}_k, \mathcal{S}_l)$. This is equivalent to $M'_j > M'_k$ where $M'_* = M_* + \sum_{i=1}^{d} right_i(\mathcal{S}_*)$. Note that this is independent from the appended seed $\mathcal{S}_l$, so we can define that $\mathcal{S}_j$ dominates $\mathcal{S}_k$, if $right_2(\mathcal{S}_j) \leq right_2(\mathcal{S}_k)$ and $M'_j \geq M'_k$.

SeqAn also implements a sparse dynamic programming algorithm by Myers and Miller (1995) with some modifications by Abouelhoda and Ohlebusch (2003) for `ChainSoP` scoring and arbitrary $d \geq 2$, see Wöhrle (2006) for more details. Note that both runtime and space requirements of this algorithm grow exponentially with respect to the seed dimension d.

$\triangleright$ SPARSECHAINING$(\mathcal{S}_1, \ldots, \mathcal{S}_n)$

1   compute top seed $\mathcal{S}_0$ and bottom seed $\mathcal{S}_{n+1}$

2    $M_0 \leftarrow 0$

3    $D \leftarrow \{\mathcal{S}_0\}$

4    $S \leftarrow \emptyset$

5   **for** $j \leftarrow 1$ **to** $n+1$ **do**

6    $\quad S \leftarrow S \cup \{\langle \mathit{left}_1(\mathcal{S}_j), j\rangle\} \cup \{\langle \mathit{right}_1(\mathcal{S}_j), j\rangle\}$

7   **for each** $\langle pos, j\rangle \in S$ in increasing order of pos **do**

8    $\quad$ **if** $pos = \mathit{left}_1(\mathcal{S}_j)$ **then**

9    $\quad\quad T_j \leftarrow \mathrm{argmax}_k \{\, \mathit{right}_2(\mathcal{S}_k) \leq \mathit{left}_2(\mathcal{S}_j) \,\big|\, \mathcal{S}_k \in D\}$ $\Big\}$ append

10    $\quad\quad M_j \leftarrow M_{T_j} + \mathit{gapscore}(\mathcal{S}_k, \mathcal{S}_j) + \mathit{weight}(\mathcal{S}_j)$ $\Big\}$ $\mathcal{S}_j$

11    $\quad$ **if** $pos = \mathit{right}_1(\mathcal{S}_j)$ **then**

12    $\quad\quad$ **if** no seed $\in D$ dominates $\mathcal{S}_k$ **then**

13    $\quad\quad\quad$ **for each** $\mathcal{S}_k \in D$ dominated by $\mathcal{S}_j$ **do**   update

14    $\quad\quad\quad\quad D \leftarrow D \setminus \{\mathcal{S}_k\}$ D

15    $\quad\quad\quad D \leftarrow D \cup \{\mathcal{S}_j\}$

16   **return** CHAINTRACEBACK$(\mathcal{S}_0, \ldots, \mathcal{S}_{n+1}, n+1, T)$

Algorithm 7: **Global Chaining by Sparse Dynamic Programming.** $\mathcal{S}_1, \ldots, \mathcal{S}_n$ is a set of 2-dimensional seeds. The algorithm may be used for gap scoring functions `Zero` and `Manhattan`, where the semantic of *dominate* depends on the scoring function.

8.6.4 Banded Alignment

Given a chain $\mathcal{C} = \langle \mathcal{S}_1, \ldots, \mathcal{S}_n \rangle$ of seeds between two sequences $a = a_1 \ldots a_n$ and $b = b_1 \ldots b_m$, we can find a good alignment using *banded alignment*. Like the alignment algorithms we described in Section 8.5, this method is based on *dynamic programming*. Remember that the Needleman-Wunsch algorithm (Section 8.5.1) computes $n \times m$ score $M_{i,j}$ of the best alignments between the prefixes $a_1 \ldots a_i$ and $b_1 \ldots b_j$. Since $\mathcal{C}$ gives us an estimate of the approximate optimal alignment, we need to compute only those values $M_{i,j}$ that lay near to $\mathcal{C}$. This *band* of width B contains the following pairs of coordinates; see Figure 28: (1) for any two characters a_x and b_y that are aligned by a seed $\mathcal{S}_k \in \mathcal{C}$ all pairs $\langle i, j \rangle$ with $|i - x| + |j - y| \leq B$, and (2) the square of pairs $\langle i, j \rangle$ with $right_1(\mathcal{S}_k) - B \leq i < left_1(\mathcal{S}_{k+1}) + B$ and $right_2(\mathcal{S}_k) - B \leq j < left_2(\mathcal{S}_{k+1}) + B$ for $k \in \{1, \ldots, n\}$. Computing only these cells of M speeds up the alignment process. SeqAn provides the function `bandedChainAlignment` that computes an optimal banded alignment following a chain.

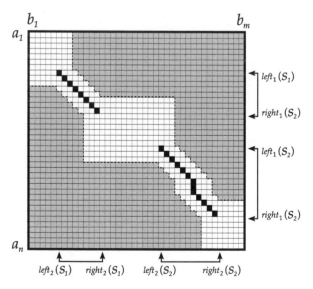

Figure 28: **Banded Alignment.** The white area of matrix M represents the *band* of width $B = 3$ around the chain $\mathcal{C} = \langle \mathcal{S}_1, \mathcal{S}_2 \rangle$.

Chapter 9

Pattern Matching

The *pattern matching* problem is to find a given needle sequence p in a haystack sequence t, for example to determine where a string t contains the string p as a substring. There are several variants of this problem:

- **Exact Matching:** Find substrings p in t. See Section 9.1 for single searching needles, and Section 9.2 for searching multiple needles.

- **Approximate Matching:** Find substrings s in t with $score(s, p) \geq k$ for a given threshold k. See Section 9.3 for alignment scoring schemes *score*.

- **Complex Pattern Searching:** p is an expression that encodes a set of strings to be found in t. See for example in Section 9.4.2 how to search for regular expressions.

In SeqAn, the function `find` finds an occurrence of a needle in a haystack; it can repeatedly be called to find all occurrences. `find` needs the following information to work: the haystack, the needle, what kind of algorithm to be used, the current state of the search (e.g., the last found position), and possibly – depending on the algorithm – some preprocessing data. This information is divided into two objects: (1) the *finder* that holds all information related to the haystack, and (2) the *pattern* that holds information related to the needle; see Figure 29.

The last found match position is stored in the finder and can be retrieved by the function `position`. This is for most searching algorithms the position of the first value of the match, except for approximate searching algorithms; since finding the begin position of an approximate match needs some additional overhead,

Online Searching

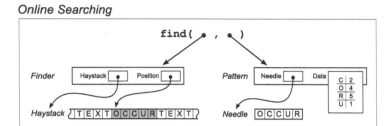

Index Searching

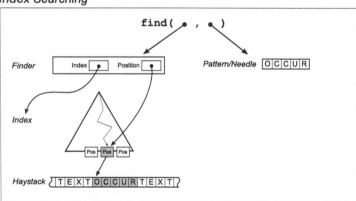

Figure 29: **Online Searching and Index Searching. Top:** Calling `find` for online searching. The search algorithm is determined by the type of the pattern, which contains all relevant preprocessing data. **Bottom:** Calling `find` for index searching (see Chapter 11). The search algorithm depends on the finder type. The needle sequence acts directly as a pattern.

the function `position` returns the position of the *last* character of the match (see Section 9.3). SeqAn also offers the functions `beginPosition` and `endPosition` for determining the begin and end position of matches explicitly, where `endPosition` is immediately available after calling `find`, and `beginPosition` requires in the case of approximate string matching a previous call of the function `findBegin` in order to find the beginning of the match. The position of a finder can also be set by the user via the function `setPosition`. If searching is started or resumed at a specific position *pos*, then only occurrences on positions $\geq pos$ will be found. Moreover, all algorithms in SeqAn guarantee that the occurrences found by calling `find` are emitted in order of increasing positions. In this chapter, we will focus on *online searching algorithms*, which solve the pattern matching problems by preprocessing the needle, not the haystack. Most online searching algorithms implemented in SeqAn are also described in (Navarro and Raffinot 2002).

9.1 Exact Searching

The *exact searching problem* is to find for a given needle $p_1 \ldots p_m$ and haystack $t_1 \ldots t_n$ all positions j for which $p_1 \ldots p_m = t_j \ldots t_{j+m-1}$. Table 18 shows some online searching algorithms for exact searching provided by SeqAn. Listing 21 shows how to use these algorithms. The performance of online algorithm depends (besides other things) on the *alphabet size* Σ and the needle length m; see Figure 31.

9.1.1 Brute-Force Exact Searching

The most simple way for searching a needle $p = p_1 \ldots p_m$ in a haystack $t = t_1 \ldots t_n$ is to compare p with $t_{pos+1} \ldots t_{pos+m}$ for each position $pos \in \{0, \ldots, n-m\}$ (Algorithm 8). This method takes time $O(n \times m)$, and it is rather slow compared to other algorithms. SIMPLESEARCH has the advantage that it is completely generic and works for arbitrary value types. All other ex-

`Simple`	A brute-force but generic searching algorithm that can deal with sequences of all value types.
`Horspool`	Horspool's algorithm (Horspool 1980) is a simple yet fast algorithm with in average sublinear searching time that is suitable for many pattern matching settings.
`ShiftOr`	An algorithm that uses bit parallelism. It should only be used for patterns that are not longer than a machine word, i.e., 32 or 64 characters. Even for small patterns, it is outperformed by Horspool for alphabets larger than `Dna`.
`BFAM`	Backward Factor Automaton Matching is an algorithm that applies an automaton of the reversed needle's substrings. It is a good choice for long patterns.
`BndmAlgo`	The Backward Nondeterministic DAWG (Directed Acyclic Word Graph) Matching algorithm uses a special automaton to scan through the haystack. It is an alternative to `BFAM` for medium-sized patterns.

Table 18: **Exact Pattern Matching Algorithms.**

```
String<char> t = "he_is_the_hero";
String<char> p = "he";
Finder<String<char> > finder(t);
Pattern<String<char>, Horspool> pattern(p);
while (find(finder, pattern))
{
    std::cout << position(finder) << ","; //output: 0,7,10
}
```

Listing 21: **Exact Online Searching Example.**

act searching algorithms in SeqAn need some additional space for storing preprocessed data, and this space could exceed memory when large alphabets are used; SIMPLESEARCH on the other hand is not limited this way, since it needs no preprocessing data at all.

▷ SIMPLESEARCH $(p_1 \ldots p_m, t_1 \ldots t_n)$

1 **for** $pos \leftarrow 0$ **to** $n - m$ **do**
2   **if** $p_1 \ldots p_m = t_{pos+1} \ldots t_{pos+m}$ **then**
3   report match at position $pos + 1$

Algorithm 8: **Brute-Force Exact Searching.**

9.1.2 Horspool's Algorithm

Horspool's algorithm (Horspool 1980) (see Algorithm 9) is a simplification of an algorithm by Boyer and Moore (1977). The algorithm compares the needle $p_1 \ldots p_m$ with the substring $t_{pos+1} \ldots t_{pos+m}$ of the haystack, where the search starts at $pos = 0$. After each comparison, pos is increased by a safe shift width k, which means that k is small enough that no possible match in between gets lost. Suppose that pos is increased by i, then t_{pos+m} will be compared to p_{m-i} during the next comparison step. Hence if $p_{m-i} \neq t_{pos+m}$ for all $1 \leq i \leq k$, then k is *safe*. The maximum safe shift width for each possible value of t_{pos+m} is stored in a preprocessed table *skip*.

The worst case running time of Horspool's algorithm is $O(n^2)$, but in practice it runs in linear or even sublinear time on average. This algorithm is a good choice for most exact pattern matching problems, except (1) if the alphabet (and hence the shift width) is very small compared to the pattern length, since in this case it is outperformed by other algorithms, or (2) for very large alphabets, since storing *skip* gets inefficient then.

Horspool's algorithm is applied when the `Horspool` specialization of the class `Pattern` is used as pattern; see Listing 21 for an ex-

```
     ▷ HORSPOOL (p₁ ... pₘ, t₁ ... tₙ)
 1     skip[c] ← m for all c ∈ Σ                        ⎫
 2     for i ← 1 to m − 1 do skip[pᵢ] ← m − i           ⎬ preprocessing
                                                        ⎭
 3     pos ← 0                                          ⎫
 4     while pos ≤ n − m do                             ⎪
 5       │ i ← m                                        ⎪
 6       │ while pᵢ = t_{pos +i} do                     ⎪
 7       │   │ if i = 1 then                            ⎬ searching
 8       │   │   └ report match at position pos + 1     ⎪
 9       │   │     break                                ⎪
10       │   └ i ← i − 1                                ⎪
11       └ pos ← pos + skip[t_{pos +m}]                 ⎭
```

Algorithm 9: **Horspool's Algorithm.**

ample.

9.1.3   Shift-Or Algorithm

Shift-Or is a simple online algorithm for exact pattern matching that benefits from bit-parallelism. For a given needle $p = p_1 \ldots p_m$ and a haystack $t_1 \ldots t_n$, we define for each $j \in \{1, \ldots, n\}$ a length-m vector b^j of boolean variables

$$b_i^j := \text{“} p_1 \ldots p_i \text{ does not match to a suffix of } t_1 \ldots t_j \text{”},$$

$i \in \{1, \ldots m\}$. If $not\ b_m^j$, then p matches t at a position $j - m + 1$. At each time j, SHIFTOR stores b^j in a bit vector b. When j is increased, b is updated according to the recursion:

$$b_i^j = b_{i-1}^{j-1}\ or\ (p_i \neq t_j).$$

SHIFTOR applies bit-parallelism, hence this takes only one left-shift operation on b and one bit-wise *or* operation with a bit vector $mask[t_j]$ (see Algorithm 10, line 5). The bit vectors $mask[c]$ are preprocessed for each possible value $c \in \Sigma$; $mask[c]_i = 0$, iff $p_i = c$.

The Shift-Or algorithm is quite fast, as long as b fits into one machine word, i.e., as long as $m <= 32$ or 64. For longer patterns,

```
    ▷ SHIFTOR (p₁ ... pₘ, t₁ ... tₙ)
1     mask[c] ← 1ᵐ for all c ∈ Σ          ⎫ preprocessing
2     for i ← 1 to m do mask[pᵢ]ᵢ ← 0     ⎭

3     b ← 1ᵐ                              ⎫
4     for j ← 1 to n do                   ⎪
5       ⌐ b ← (b << 1)| mask[tⱼ]          ⎬ searching
6       | if bₘ = 0 then                  ⎪
7       ⌐ ⌐ report a match at j − m + 1   ⎭
```

Algorithm 10: **Shift-Or Algorithm.**

multiple machine words must be used, but this diminishes the positive effect of the bit-parallelism. Moreover, Shift-Or is outperformed by Horspool's algorithm (Section 9.1.2) for all but very small alphabets (see Figure 30 on page 145). Shift-Or is therefore best for small patterns and small alphabets.

9.1.4   Backward Factor Automaton Matching

Backward Factor Automaton Matching (BFAM) is an exact online algorithm that applies an automaton, e.g., an *oracle automaton* as described by Allauzen, Crochemore, and Raffinot (2001) or a *trie*. The principle of BFAM is *Backward Factor Searching* as presented in Algorithm 11: BF reads a suffix of $t_1 \ldots t_{pos+m}$ from back to front until either a match of p is found, or p does not contain any substring (*factor*) that matches to the read suffix. If p does not contain $t_{pos+k} \ldots t_{pos+m}$, then k is a *safe shift*, i.e., *pos* can be increased by k without losing a match, since any substring of t that starts at a position between $pos+1$ to $pos+k$ also contains $t_{pos+k} \ldots t_{pos+m}$.

The main question in BF is how to check the condition in line 4 whether p contains $t_{pos+k} \ldots t_{pos+m}$. For that purpose, BFAM (Algorithm 12) applies a *factor automaton* on the reverse needle $p_m \ldots p_1$, i.e., an automaton that accepts all substrings of this sequence. This automaton is processed on $t_{pos+m} \ldots t_{pos+1}$ until either a match is found, or an undefined state is reached because the needle does not contain the string $t_{pos+k} \ldots t_{pos+m}$.

$\triangleright$ BF $(p = p_1 \ldots p_m, t_1 \ldots t_n)$

1 $pos \leftarrow 0$
2 **while** $pos \leq n - m$ **do**
3    $k \leftarrow m$
4   **while** p contains $t_{pos+k} \ldots t_{pos+m}$
        do
5   **if** $k = 1$ **then**
6   report match at $pos+1$
7   **break**
8      $k \leftarrow k - 1$
9    $pos \leftarrow pos+k$

Algorithm 11: **Backward Factor Searching Principle.**

In SeqAn, the kind of automaton is specified when choosing the specialization of the class `Pattern`:   The specialization `BFAM<Oracle>` is used for applying an oracle automaton (see Section 12.1.2) and `BFAM<Trie>` for a suffix trie (see Section 12.1.1). Oracle automata may also accept strings other than substrings of $p_m \ldots p_1$, and this may lead to shorter shift widths. Fortunately, the only length-m string accepted by the oracle is $p_m \ldots p_1$ itself, so we need no additional verification in line 6 of Algorithm 11, as it will be necessary for MULTIBFAM in line 15 of Algorithm 15. Oracles are more compact than suffix tries: The oracle of $p_m \ldots p_1$ has only $m + 1$ states and at most $2 \times m$ transitions, whereas the number of states and transitions of a suffix trie can be quadratic. This parsimony benefits the run time, because a smaller automaton has better chances to stay in cache, and because oracles take less time to be built up. A comparison between the run times of the two variants (Figure 30 on page 145) reveals that BFAM<TRIE> is slightly faster than BFAM<ORACLE> for small alphabets and needle lengths, whereas for large alphabets or needle lengths the oracle takes advantage of its space efficiency.

9.1.5 Backward Nondeterministic DAWG Matching

The *BNDM algorithm (Backward Nondeterministic Directed Acyclic Word Graph Matching)* is a bit-parallel variant of an al-

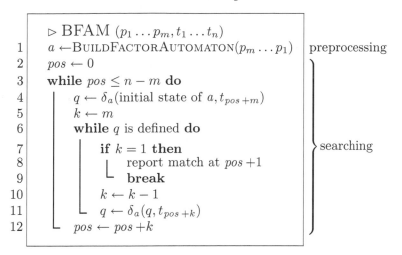

	$\triangleright$ BFAM $(p_1 \dots p_m, t_1 \dots t_n)$	
1	$a \leftarrow$ BUILDFACTORAUTOMATON$(p_m \dots p_1)$	preprocessing
2	$pos \leftarrow 0$	
3	**while** $pos \leq n - m$ **do**	
4	$\quad q \leftarrow \delta_a(\text{initial state of } a, t_{pos+m})$	
5	$\quad k \leftarrow m$	
6	$\quad$ **while** q is defined **do**	searching
7	$\quad\quad$ **if** $k = 1$ **then**	
8	$\quad\quad\quad$ report match at $pos+1$	
9	$\quad\quad\quad$ **break**	
10	$\quad\quad k \leftarrow k - 1$	
11	$\quad\quad q \leftarrow \delta_a(q, t_{pos+k})$	
12	$\quad pos \leftarrow pos + k$	

Algorithm 12: **Backward Factor Automaton Searching.** δ_a is the transition function of a. Note that a is built on the *reverse* needle $p_m \dots p_1$.

gorithm by Crochemore et al. (1994). It applies bit-parallelism for tracking the substrings of the needle during backward factor searching (BF, see Algorithm 11, line 4). Let us define for each $k \in \{1, \dots, m\}$ a length-m vector b^k of boolean variables

$$b_i^k := \text{``}t_{pos+k} \dots t_{pos+m} \text{ matches a prefix of } p_i \dots p_m\text{''}.$$

BNDM (Algorithm 13) stores b^k in a bit vector b, which is updated when k is decreased according to the recursion:

$$b_i^k = b_{i+1}^{k+1} \text{ and } (p_i = t_{pos+k}).$$

This takes two bit-parallel operations: One right shift in line 14 and one bit-wise *and* with a preprocessed bit vector $mask[t_{pos+i}]$ in line 9 of Algorithm 13. The bit vectors $mask[c]$ are preprocessed for each possible value $c \in \Sigma$; $mask[c]_i = 1$, iff $p_i = c$.

BNDM improves the *safe* shift width of BF as follows: If $b_1^k = 0$, then $t_{pos+k} \dots t_{pos+m}$ does not match to a prefix of p, hence p does not match t at position $pos+k$. Suppose that k is a *safe* shift, then $k+1$ will also be *safe*. Hence, we need only to take into account shift widths k with $b_1^k = 1$. The variable *skip* stores the last found k for which $b_i = 1$ (line 12). *skip* is then used in line 15 as shift width.

$\triangleright$ BNDM $(p_1 \ldots p_m, t_1 \ldots t_n)$

```
1    mask[c] ← 0^m for all c ∈ Σ          ⎫ preprocessing
2    for i ← 1 to m do mask[p_i]_i ← 1     ⎭
3    pos ← 0
4    while pos ≤ n − m do
5    │   k     ← m
6    │   skip  ← m
7    │   b     ← 1^m
8    │   while b ≠ 0^m do                   searching
9    │   │   b ← b ∧ mask[t_{pos + k}]
10   │   │   k ← k − 1
11   │   │   if b_1 = 1 then
12   │   │   │   if k > 0 then skip ← k
13   │   │   └   else report match at pos +1
14   │   └   b ← b >> 1
15   └   pos ← pos + skip
```

Algorithm 13: **Backward Nondeterministic DAWG Matching.**

9.1.6 Results

Figure 30 shows the average run times divided by the length of the searched haystack for searching needles of length m. The machine word size was 32, so the run times of the bit parallel algorithm `ShiftOr` and `BndmAlgo` are discontinuous at multitudes of 32. For small alphabets like DNA ($|\Sigma| = 4$), the fastest algorithms are `ShiftOr` for small m, `BFAM<Trie>` for middle sized m, and `BFAM<Oracle>` for large m. For larger alphabets, e.g., when searching proteins or English texts, either `Horspool` for small m or `BFAM` for larger m is the fastest; see Figure 31. Compared with the results of Navarro and Raffinot (2002, Fig. 2.22), our implementation of the BNDM algorithm is outperformed in any parameter setting.

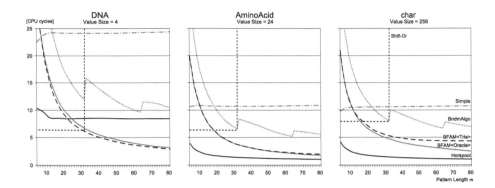

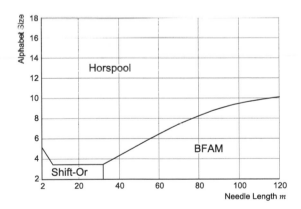

Figure 30: **Run Times of Exact Pattern Matching Algorithms.** The average run times per haystack value for different exact pattern matching algorithms depending on the length m of the needle. We searched for patterns in (1) the genome of *Escherichia coli*, (2) proteins from the Swiss-Prot database, and (3) the English Bible.

Figure 31: **Fastest Exact Pattern Matching Algorithm.** The best algorithm for searching all exact occurrences of length-m patterns in random haystacks, depending on the size $|\Sigma|$ of the alphabet. Since the pattern is a random string, `Horspool` gains ground compared to Figure 30.

9.2   Exact Searching of Multiple Needles

In this section, we describe algorithms that search several needles $p^1, \ldots, p^k$ in a haystack $t_1 \ldots t_n$ at once, which is in general faster than searching one needle after the other. We search the pairs (i, j) for which p^i matches a prefix of $t_j \ldots t_n$. Many algorithms proposed in (Navarro and Raffinot 2002) are implemented in SeqAn; see Table 19. Here, we will only describe two of them in more details, since these two outperform all others in almost any case. An comparison of run times can be found in Figure 32 on page 150.

Listing 22 shows an example of how to use the function `find` to search for multiple patterns at once. After each call, `position(pattern)` returns the index number of the needle that was recently found starting at `position(finder)` in the haystack. Some algorithms ensure that hits emerge in a certain order. For example, the hits found by `WuManber` and `MultiBFAM` are sorted in increasing order by `position(finder)`, and two hits at the same position `position(finder)` are sorted in increasing order by `position(pattern)`. The example program therefore finds first `"the"` at position 0, then `"hero"` at position 4, then `"theory"` at position 11, and finally `"the"`, also at position 11.

9.2.1   Wu-Manber Algorithm

The algorithm by Wu and Manber (1994) is an extension of Horspool's algorithm (see Section 9.1.2, page 139) for multiple needles. WuMANBER (Algorithm 14) compares the needles $p^1, \ldots, p^k$ to the haystack t at a position pos, which is then increased by a *safe shift width*. For multiple needles, it is not advisable to select the shift width depending on the occurrences of a *single* value t_{pos+m} within the needles, as it is done by HORSPOOL (Algorithm 9, page 140), since this would lead to rather small shift widths and hence a poor performance, because it is rather probable to find each possible value in the ending region of at least one of the needles. WuMANBER uses therefore $q \geq 2$ values

WuManber	An extension of Horspool's algorithm (Horspool 1980) for searching multiple needles. The favored algorithm in many cases.
MultiBFAM	Backward Factor Automaton Matching for Multiple sequences is an extension of BFAM for searching multiple sequences. It applies an automaton that accepts the reversed substrings of the needles' prefixes. This algorithm is a good choice for long patterns, small alphabets, or large needle sets.
AhoCorasick	An algorithm by Aho and Corasick (1975) that uses an extended trie automaton to scan through the haystack sequence. It performs well especially for small alphabets pattern lengths.
MultipleShiftAnd	An extension of the Shift-And algorithm for multiple patterns. This algorithm is competitive only if the sum of the needle lengths is smaller than the size of one machine word.
SetHorspool	Another extension of Horspool's algorithm (Horspool 1980) for multi-pattern searching that applies a trie of the reverse needles. In practice, it is outperformed by WuManber.

Table 19: **Exact Pattern Matching Algorithms for Searching Multiple Needles.**

```
String<char> t = "the_heroes_theory";
String<String<char> > p;
appendValue(p, "theory");
appendValue(p, "hero");
appendValue(p, "the");
Finder<String<char> > finder(t);
Pattern<String<String<char> >, WuManber> pattern(p);
while (find(finder, pattern))
{
   std::cout << "found pattern " << position(pattern)
             << "at position "   << position(finder)
             << ", ";
}
```

Listing 22: **Multiple Pattern Searching Example.**

$t_{pos+m-q+1} \ldots t_{pos+m}$ for determining the shift widths. A preprocessed table *shift* stores for each q-gram $w \in \Sigma^q$ a safe shift width. *shift* may use hashing if a table size $|\Sigma|^q$ would be too large for storing a shift width for each q-gram in memory. A second table *verify* is used to determine which needles possibly match and are verified.

The expected shift widths are optimal if q is selected such that the number $|\Sigma|^q$ of possible q-grams is about the same as the number of overlapping q-grams occurring in needles. In practice, the optimal q may be smaller, because the computation of a hash value needed to access *shift* and *verify* takes time $O(q)$, and this dominates the performance of the main loop (lines 11 to 18).

9.2.2   Multiple BFAM Algorithm

The Multiple Backward Factor Automaton Matching algorithm extends the BFAM algorithm (see Section 9.1.4, page 141 and Algorithm 15 on page 151) for searching multiple needles. The factor automaton, e.g., a factor oracle (Section 12.1.2), is built for the reverse needles $p^1, \ldots, p^k$. Since the maximal safe shift width cannot be larger than the length m of shortest needle, the automaton considers only the prefixes of the needles that do not

```
   ▷ WUMANBER (P = {p¹, ..., pᵏ}, t₁ ... tₙ)
 1    m ← minimum length of pʲ
 2    z ← m − q + 1
 3    shift[w] ← z for all w ∈ Σ^q                    ⎫
 4    for i ← 1 to z do                               ⎪
 5        for j ← 1 to k do                           ⎬ build shift
 6            shift[pⁱⱼ ... pʲᵢ₊q₋₁] ← z − i         ⎭
 7    verify[w] ← {} for all w ∈ Σ^q                  ⎫
 8    for j ← 1 to k do                               ⎬ build verify
 9        verify[pʲz ... pʲₘ] ← verify[pʲz ... pʲₘ] ∪ {j}  ⎭
10    pos ← 0                                         ⎫
11    while pos ≤ n − m do                            ⎪
12        w ← t_{pos+z} ... t_{pos+m}                 ⎪
13        if shift[w] = 0 then                        ⎪
14            for each j ∈ verify[w] do               ⎬ searching
15                report if pʲ matches t at pos +1    ⎪
16            pos ← pos +1                            ⎪
17        else                                        ⎪
18            pos ← pos + shift[w]                    ⎭
```

Algorithm 14: **Wu-Manber Searching of Multiple Needles.**

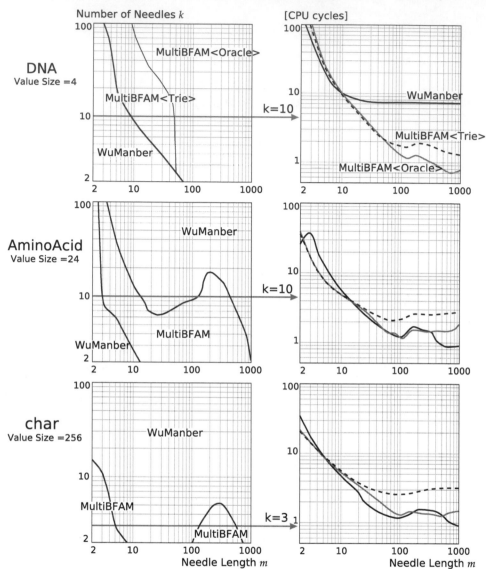

Figure 32: **Fastest Multiple Pattern Matching Algorithm. Left:** The optimal algorithm for finding k patterns of length m in parts of (1) the *Escherichia coli* genome (2) proteins from the Swiss-Prot database, and (3) the English Bible. If $|\Sigma = 4|$, then WuManber is optimal for small m and MultiBFAM for large m. For larger alphabets, the reverse is true. **Right:** Slices through the left figures show the actual run time per searched haystack value for searching $k = 10$ (Dna or AminoAcid) or $k = 3$ (char) patterns. Both algorithms are sublinear, so for large m the run times may even fall below 1 CPU cycle per haystack value.

exceed that length. During the search, the automaton processes a part $t_{pos+1} \ldots t_{pos+m}$ of the haystack from back to front. If the whole substring can be processed, MULTIBFAM tests all needles in $verify[q]$, which gives for the current automaton state q the list of needles with prefix $t_{pos+1} \ldots t_{pos+m}$.

$\triangleright$ MULTIBFAM $(P = \{p^1, \ldots, p^k\}, t_1 \ldots t_n)$

1   $m \leftarrow$ minimum length of p^j

2   $rev^j \leftarrow p^j_m \ldots p^j_1$ for $j \in \{1, \ldots, k\}$

3   $a \leftarrow$ BUILDFACTORAUTOMATON$(rev^1, \ldots, rev^k)$ } build a

4   $verify[q] \leftarrow \{\}$ for all states q in a

5   **for** $j \leftarrow 1$ **to** k **do**

6      $q \leftarrow$ the state of a after processing rev^j

7      $verify[q] \leftarrow verify[q] \cup \{j\}$ } build $verify$

8   $pos \leftarrow 0$

9   **while** $pos \le n - m$ **do**

10      $q \leftarrow \delta_a($initial state of $a, t_{pos+m})$

11      $k \leftarrow m$

12   **while** q is defined **do**

13   **if** $k = 1$ **then**

14   **for each** $i \in verify[q]$ **do**

15   report if p^i matches t at $pos+1$

16   **break**

17          $k \leftarrow k - 1$

18          $q \leftarrow \delta_a(q, t_{pos+k})$

19      $pos \leftarrow pos + k$ } searching

Algorithm 15: **Backward Factor Automaton Matching for Multiple Needles.**

9.3   Approximate Searching

So far, we discussed *exact* matching, that means a match of the search is a substring s of the haystack t that equals the pattern *p*. In this section, we will relax this condition, such that a match *s* needs not be equal but only sufficiently *similar* to *p*. More precisely spoken, we want to find all substrings *s* of *t* for which the distance $dist(s, p)$ with respect to a certain distance metric *dist* does not exceed a certain threshold *T*. This is called *approximate string matching*. In its most general form, *dist* could be any sequence distance measure based on alignment scores as described in Section 8.3.1, though most approximate search algorithms are specialized for *edit distance*. SeqAn also supports approximate search algorithms that only allow mismatches between *s* and *p* but no inserts or deletes; we will describe them in Section 9.4.1.

When an approximate searching algorithm searches the haystack *t* starting from a position *pos*, then we can either search for matching substrings of $t_{pos} \ldots t_n$ (infix search), or for matching prefixes of $t_{pos} \ldots t_n$ (prefix search). We will focus on infix search in this section. SeqAn offers two algorithms for prefix search, namely the algorithms by Sellers and Myers that can be easily adapted for prefix search.

In SeqAn, finding all infix hits *s* in *t* is done in two steps:

(1) The function `find` looks for a position in *t* at which a match ends. The threshold *T* is either set by calling `setScoreLimit` or simply passed to `find` as a third function argument.

(2) If a match was found, the function `findBegin` can be used to search its begin position. The threshold *T* for that approximate search can be passed as a third argument, otherwise the function applies the same threshold as during the last call of `find`. Subsequent calls of `findBegin` may be used to find several begin positions to the same end position.

Technically, `findBegin` is implemented as a prefix search on the *reverse* needle and haystack strings. Listing 23 demonstrates how to use `find` and `findBegin`.

The algorithms for approximate string matching supported by SeqAn are listed in Table 20. They are also described in the book of Navarro and Raffinot (2002). Another good survey of approximate string matching algorithms can be found in (Navarro 2001).

```
String<char> t = "babybanana";
String<char> p = "babana";
Finder<String<char> > finder(t);
Pattern<String<char>, Myers<FindInfix> > pattern(p);
while (find(finder, pattern, -2))
{
   std::cout << "end: " << endPosition(finder) << std::endl;
   while (findBegin(finder, pattern, getScore(pattern)))
   {
      std::cout << "begin: " << beginPosition(finder)
                << std::endl;
      std::cout << infix(finder) << " matches with score "
                << getBeginScore(pattern) << std::endl;
   }
}
```

Listing 23: **Approximate String Searching Example.** The program finds six substrings "babyba", "byban", "bybana", "banan", "bybanan", and "banana" of the haystack "babybanana" that match the needle "babana" with at least two errors (edit distance). Note that the two matches "banan" and "bybanan" both end at the same position 9. The third argument of findBegin is optional; the default is the score limit T of the last call of find, i.e., -2 in this example. If we use this, six more matches would be found.

DPSearch	An algorithm by Sellers (1980) that is based on the dynamic programming algorithm for sequence alignment by Needleman and Wunsch (1970). It can also be used for prefix search.
Myers	A fast searching algorithm for edit distance using bit parallelism by Myers (1999). It can also be used for prefix search.
Pex	A filtering technique by Navarro and Baeza-Yates (1999) that splits the needle into $k + 1$ pieces and search these pieces exactly in the haystack.
AbndmAlgo	Approximate Backward Nondeterministic DAWG Matching, an adaption of the BNDM algorithm for approximate string matching.

Table 20: **Approximate Pattern Matching.** Specializations of Pattern.

9.3.1   Sellers' Algorithm

The algorithm by Sellers (1980) resembles the dynamic programming alignment algorithm (Needleman and Wunsch 1970) with free start gaps for the haystack t, as it was described in Section 8.5.4. Remember that FILLMATRIX$(p_1 \ldots p_m, t_1 \ldots t_n)$ (see Algorithm 2 on page 120) uses a matrix M, where $M_{i,j}$ is the optimal score for aligning the two prefixes $p_1 \ldots p_i$ and $t_1 \ldots t_j$. If we initialize $M_{0,j} \leftarrow 0$ for all $j \in \{1, \ldots, n\}$, then each $M_{i,j}$ is filled with the optimal score of alignments between $p_1 \ldots p_i$ and a suffix of $t_1 \ldots t_j$.

The j-th pass of the outer loop in SELLERS (Algorithm 16) computes the j-th column C of the matrix M. The inner loop computes in line 7 the value $C_i = M_{i,j}$ according to Equation 8.6 (page 118). The variable d was previously set to $M_{i-1,j-1}$ (case 1), v to $M_{i-1,j}$ (case 2), and h to $M_{i,j-1}$ (case 3). At the end of the inner loop (line 10), the value $v = C_m = M_{m,j}$ is the optimal score of an alignment between $p = p_1 \ldots p_m$ and a suffix s of $t_1 \ldots t_j$. If the sequence distance $-v$ between s and p is $\leq T$, then s is an approximate match and its end position j is reported.

$$
\begin{array}{ll}
& \triangleright \text{SELLERS } (p_1 \ldots p_m, t_1 \ldots t_n, T) \\
1 & C_i \leftarrow i \times g \text{ for each } i \in \{1, \ldots, m\} \\
2 & \textbf{for } j \leftarrow 1 \textbf{ to } n \textbf{ do} \\
3 & \quad v \leftarrow 0 \\
4 & \quad d \leftarrow 0 \\
5 & \quad \textbf{for } i \leftarrow 1 \textbf{ to } m \textbf{ do} \\
6 & \quad\quad h \leftarrow C_i \\
7 & \quad\quad v \leftarrow \max\{d + \alpha(p_i, t_j), \max\{v, h\} + g\} \\
8 & \quad\quad C_i \leftarrow v \\
9 & \quad\quad d \leftarrow h \\
10 & \quad \textbf{if } -v \leq T \textbf{ then } \text{report match end position } j
\end{array}
$$

Compute in C the j-th column of the matrix M

Algorithm 16: **Sellers' Algorithm.** α returns the score of aligning two values; g is the (usual negative) gap score.

SELLERS can easily be extended to support *affine gap costs* following Gotoh's idea (Gotoh 1982), which we described in Section 8.5.2. SeqAn supports both variants for linear and for non-linear gap costs, and selects it according to the applied scoring scheme.

The algorithm can also be adapted for *prefix searching*, we just have to change the initialization to make start gaps in the text non-free. That is, we change lines 3 and 4 in SELLERS to $v \leftarrow i \times g$ and $d \leftarrow (i-1) \times g$.

Ukkonen's Trick

Sellers' algorithm takes time $O(nm)$ for finding the occurrences of a pattern $p_1 \ldots p_m$ in a text $t_1 \ldots t_n$. With a slight modification (Ukkonen 1985), this can be accelerated for *edit distance* scoring to $O(km)$ on average, where $k = -T$ is the number of allowed errors per match. The *trick* is to compute just the cells C_i for $i \leq i_0$, where i_0 is minimal such that $C_{i'} < T$ for all $i' > i_0$. At the beginning of SELLERS C_i is initialized to $C_i = ig = -i$, so we set $i_0 \leftarrow -T$. Suppose that we know an i_0 for a given column j; one can easily prove that for the next column $j+1$ holds $C_{i'} < T$ for all $i' > i_0 + 1$, i.e., the i_0 must be increased by at most one. After computing the values $C_0, C_1, \ldots, C_{i_0}, C_{i_0+1}$, we can easily calculate the actual i_0 in (amortized) constant time.

9.3.2   Myers' Bitvector Algorithm

Myers (1999) uses bit parallelism to speed up Sellers' algorithm for *edit distance*.[1] Remember that the edit distance between two sequences is the negative score of their optimal alignment where each match scores 0 and each mismatch and gap scores -1 (see Section 8.3.1). The main idea of this algorithm is to encode the j-th column of the matrix M of the Needleman-Wunsch algorithm (Algorithm 2) in five bit vectors, each of length m (the length of

---

[1] We present here the variant by Hyyro and Fi (2001).

the needle):

$$VP_i := (M_{i,j} = M_{i-1,j} - 1) \qquad VN_i := (M_{i,j} = M_{i-1,j} + 1)$$
$$HP_i := (M_{i,j} = M_{i,j-1} - 1) \qquad HN_i := (M_{i,j} = M_{i,j-1} + 1)$$
$$D0_i := (M_{i,j} = M_{i-1,j-1})$$

where $i \in \{1, \ldots, m\}$. In each pass of the main loop, MYERS (Algorithm 17) computes these five vectors for column j based on of the vectors for column $j-1$, which takes 15 bit vector operations. If the bit vectors do not fit into one machine word, then several machine words per bit vector must be used to store VP and VN – all other bit vectors need not be stored completely. MYERS also keeps track of the current score *score* and reports the position j whenever it climbs above the (negative) score limit T, i.e., the number of errors falls below $-T$.

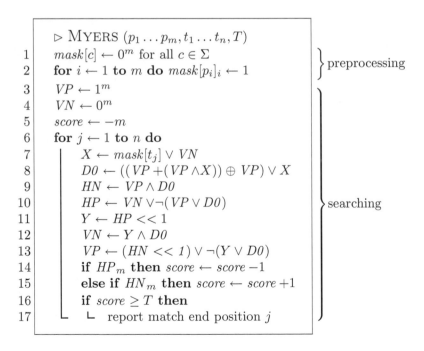

Algorithm 17: **Myers' Bit-vector Algorithm.**

MYERS would perform a prefix search if we force $M_{0,j} = M_{0,j-1} - 1$, i.e., $HP_0 = 1$. This can simply be done by changing line 11 to $Y \leftarrow HP << 1|1$.

SeqAn contains two implementations of Myers' algorithm, one for needle length up to one machine word, and a second for longer needles. The algorithm is the fastest approximate searching function in SeqAn for high edit distances (e.g., < 60% identity; see Figure 33).

We combined Myers' algorithm with Ukkonen's Trick (Section 9.3.1), so the algorithm will usually compute only the first machine word of the bit vectors at all positions in the haystack but those at which it has regions very similar to the pattern. This makes the average running time roughly independent from the pattern length m; see Figure 33.

9.3.3 Partition Filtering

In this section, we will discuss an algorithm by Navarro and Baeza-Yates (1999) that is based on a simple idea proposed by Wu and Manber (1992): Let T be the threshold for the edit distance score when searching the needle p in the haystack t, that is we want to find all occurrences of p in t with $\leq k = -T$ errors. If we cut p into $k + 1$ pieces, then each approximate match of p in t must contain at least one of these pieces unchanged. So we start with an exact multi-pattern search for the pieces and then *verify* each occurrence of one of these pieces, that is we try to find p in the neighborhood of the found piece; see Algorithm 18. The following lemma guarantees that this approach works:

(Pidgeonhole Principle)

Let $a^1 \ldots a^l = a$ be a partition of the string a into m substrings, and let $r^1, \ldots, r^l$ be l positive integer numbers. If a matches to a string b with less than $r^1 + r^2 + \ldots + r^l$ errors, then at least one a^i matches with less than r^i errors to a substring of b.

The lemma follows directly from the fact that, for linear gap costs, the score of an alignment is the sum of the scores of its pieces. Hence, if each a^i matches to its counterpart in b with $\geq r^i$ errors, then a and b would match only with $\geq r^1 + r^2 + \ldots + r^m$ errors.

In our case, we use the lemma with $l = k + 1$ and $r^1 = r^2 = \ldots =$

$r^l = 1$.

$$
\begin{array}{ll}
 & \rhd \text{PEX } (p = p_1 \ldots p_m, t = t_1 \ldots t_n, -k) \\
1 & \text{divide } p \text{ into } k+1 \text{ parts } p^1 \ldots p^{k+1} \\
2 & pieces \leftarrow \text{FINDEXACT}(\{p^1, \ldots, p^{k+1}\}, t) \\
3 & \textbf{for each } (i, pos) \in pieces \textbf{ do} \\
4 & \quad \text{let } p^i = p_l \ldots p_{l+m} \\
5 & \quad hits \leftarrow \text{FINDAPPROX}(p, t_{pos-l-k} \ldots t_{pos+m-k}, -k) \\
6 & \quad \text{report all matches in } hits
\end{array}
$$

$\left.\begin{array}{}\\\\\end{array}\right\}$ verification

Algorithm 18: **Partition Filtering Algorithm.** FINDEXACT could be any exact searching algorithm for multiple needles, e.g., WUMANBER or MULTIBFAM, and FINDAPPROX any other stand-alone approximate string matching algorithm, e.g., SELLERS or MYERS.

We call this technique *filtering*, since all parts of the haystack that do not contain any piece of the pattern are *filtered out* and only the rest is passed to the verification process. This works well for long patterns and small error rates, since in this case the pieces are relatively long, so the expected false positive rate of the filtering is low. A good partitioning strategy is to split the needle $p_1 \ldots p_m$ into $k+1$ pieces of approximately equal length $\geq \lfloor m/(k+1) \rfloor$.

On the other hand, the costs for each verification move up for large patterns. For that reason, Navarro and Baeza-Yates (1999) applied an optimization called hierarchical verification: Let $\mathcal{T}$ be a tree with $k+1$ leafs that are labeled from p^1 to p^{k+1}. We use a balanced binary tree, but the idea works for any tree topology. We label each vertex v of $\mathcal{T}$ with the concatenated pieces of p on the leafs of the subtree rooted in v, and we call this label $p(v)$ and the number of leafs in this subtree $r(v)$. The root of $\mathcal{T}$ is therefore labeled with p. Suppose that we find in the haystack at position *pos* an exact match of p^i, then we follow the path from the i-th leaf to the root. At each vertex v of this path, we search $p(v)$ in the neighborhood of *pos* with less than $r(v)$ errors. If no occurrence of $p(v)$ is found, the verification stops, otherwise we proceed with parent of v in $\mathcal{T}$ until the root has been reached and a match of p

was verified.

It is easy to prove that this kind of verification is correct, i.e., no approximate match gets lost: Let $v_1, \ldots, v_l$ be the children of the root, then $r(v_1) + \ldots + r(v_l) = k + 1$. Hence, according to Lemma 9.3.3, any approximate match of p with at most k errors also contains some $p(v^i)$ with less than $r(v^i)$ errors. The same argument can recursively be applied to v^i, then to one of its children, and so forth.

9.4 Other Pattern Matching Problems

There are many more variants of pattern matching problems, and SeqAn provides algorithms for some of them. In this section, we will describe two of them: (1) the *k-mismatch problem* and (2) searching with *wildcards*.

9.4.1 *k*-Mismatch Searching

A *mismatch* between a sequence $a = a_1 \ldots a_n$ and another sequence $b = b_1 \ldots b_n$ is a position $i \in \{1, \ldots, n\}$ such that $a_i \neq b_i$. The number of mismatches between a and b is called the Hamming distance of the two sequences. Given a needle $p_1 \ldots p_m$ and a haystack $t_1 \ldots t_n$, then *searching with k mismatches* means to find all substrings $t_{pos+1} \ldots t_{pos+m}$ that have Hamming distance to $p_1 \ldots p_m$ of $\leq k$. This kind of searching resembles approximate string matching as described in Section 9.3 but without inserts or deletes, gaps are forbidden. Sellers' algorithm (Section 9.3.1) can be used for searching with mismatches if the costs for gaps are set to $+\infty$; but of course there are also algorithms especially for the k-mismatch problem. SeqAn for example offers the specialization `HammingHorspool` of `Pattern`, that implements an adaption of the exact pattern matching algorithm (Section 9.1.2) proposed by Tarhio and Ukkonen (1990).

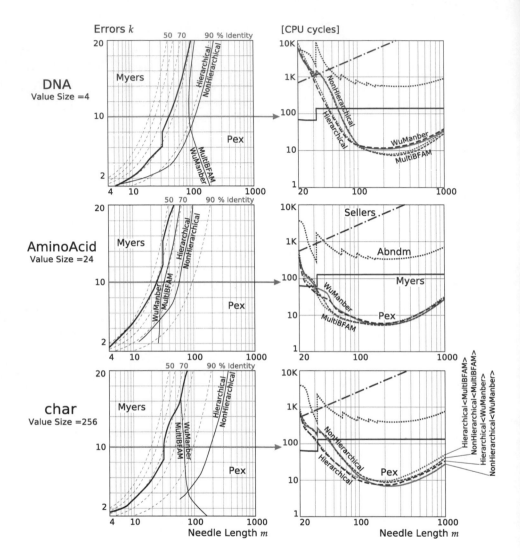

Figure 33: **Fastest Approximate Matching Algorithm. Left:** The optimal algorithm for finding a pattern of length m with at most k errors (*edit distance*) in parts of (1) the *Escherichia coli* genome (2) proteins from the Swiss-Prot database, and (3) the English Bible. **Right:** Slices through the left figures show the actual run time per searched haystack value for $k = 10$.

9.4.2 Searching with Wildcards

SeqAn provides the algorithm `WildShifAnd` that is able to search for regular expressions (e.g., Navarro and Raffinot (2002), Section 4.5.1). Note that this algorithm does not support the complete functionality of usual regular expressions, e.g., it does not support alternatives, and quantifiers like *, +, ? and {i, j} can refer only to single characters or character classes. A pattern for searching in texts $t_1 \ldots t_n$ of the alphabet Σ is a string $c^1 \ldots c^m$ that is the concatenation of *clauses*, where each clause c^i has one of the forms in Table 21. For example, the pattern

$$[A - Z0 - 9].*$$

matches all strings that start with a capital letter or a digit.

Clause	Description
w	a string
.	any character
$a\{i, j\}$	repeat a at least i and at most j times
$a*$	repeat a for 0, 1, or more times
$a+$	repeat a for 1 or more times
$a?$	optional a, same as $a\{0, 1\}$
$[b^1 \ldots b^k]$	a character in a class: Each b^j is either a set of characters from Σ, or it has the form a_1-a_2 which denotes all characters in Σ between a_1 and a_2.

Table 21: **Regular Expression Syntax.** A regular expression may contain several clauses. w is a string of values from Σ, and a is either a single character from Σ or a character class $[b^1 \ldots b^k]$.

Chapter 10

Motif Finding

Motif finding means to find matching substrings of $d \geq 2$ given sequences $a^1, \ldots, a^d$, where *matching* can either be meant to be *exact*, i.e., the matching substrings must be exactly the same, or *approximate*, i.e., differences between the substrings are allowed. The matching substrings are called motifs. In the following, we will first concentrate on the *pairwise motif finding* problem, that is finding motifs in $d = 2$ sequences. In contrast to *pattern matching* (see Chapter 9), where we search for a *complete* needle sequence in a haystack sequence, (pairwise) motif addresses the problem of finding *parts* of the needle within the haystack. In most cases, we are only interested in motifs that fulfill certain criteria of quality, for example a minimal length, a minimum alignment score or – in the case of approximate motif finding – a maximum mismatch count.

SeqAn offers algorithms for solving various kinds of motif finding problems which are spread over several modules of the library:

- **Local Alignments**: Algorithms for solving the global alignment problem (see Section 8.5), that is to find an optimal alignment between two *complete* sequences, can be adapted for motif finding, that is to find optimal alignments between *substrings* of two sequences. This is called local alignment, and we describe it in Section 10.1. The motifs found by this search are local alignments stored in alignment data structures; see Section 8.2.

- **Index Iterators**: Some index data structures, as for example the *suffix trees* or the *enhanced suffix arrays* (ESA), can be used to find exact matches between two or more sequences. SeqAn offers some special iterators (see Section 11.4.2) that can be used to browse through all exact motifs, which are defined by the begin positions and the length of the matching substrings.

- **Seed Based Motif Search**: Algorithms for expanding and combining small motifs (so-called *seeds*) to larger motifs are introduced in the Section 10.2. Seeds are represented essentially by the begin and end positions of the matching substrings, which can be stored in a data structure called Seed; see Section 8.6.1.

- **Multiple Sequence Motifs**: Algorithms for finding subtle motifs of fixed length in multiple sequences are discussed in Section 10.3. Motifs of this kind are either represented by a consensus sequence or by a position dependent weight matrix.

10.1   Local Alignments

10.1.1   Smith-Waterman Algorithm

Smith and Waterman (1981) adapted the Needleman-Wunsch algorithm (Needleman and Wunsch (1970); see Section 8.5.1), for finding motifs in two sequences a and b: The algorithm finds a substring a' of a and a substring b' of b and an alignment $\mathcal{A}$ between a' and b', such that the score of $\mathcal{A}$ is at least as good as the score of any other alignment between a substring of a and a substring of b. In this case, we call $\mathcal{A}$ an *optimal* local alignment between a and b.

The Smith-Waterman algorithm works as follows: SMITHWATERMAN (Algorithm 19) computes – just like NEEDLEMANWUNSCH (Algorithm 2) on page 120 – an $m \times n$ matrix M of scores, but other than the Needleman-Wunsch algorithm that set $M_{i,j}$ to the score of the optimal alignment between $a_1 \ldots a_i$ and $b_1 \ldots b_j$, the Smith-Waterman algorithm computes instead the optimal score of any alignment between a *suffix* of $a_1 \ldots a_i$ and a *suffix* of $b_1 \ldots b_j$.

Let these suffixes be a' and b' for a given i and j. Note that a' or b' or both could be the empty string ϵ. If both $a' = b' = \epsilon$, then $M_{i,j} = score(\epsilon, \epsilon) = 0$. Otherwise, $M_{i,j}$ can be computed by recursion (8.6) on page 118. If either $i = 0$ or $j = 0$, then obviously $M_{i,j} = 0$. After filling the complete matrix M, we get an optimal local alignment by starting a trace back at a cell $M_{i,j}$ with a maximal value, which is the score of the optimal local alignment.

```
Align< String<char> > ali;
appendValue(rows(ali), "aphilologicaltheorem");
appendValue(rows(ali), "bizarreamphibology");
Score<int> scoring(3,-3,-2);
int score = localAlignment(ali, scoring, SmithWaterman());
cout << ali;
```

Listing 24: **Smith-Waterman Algorithm.** Finding an optimal local alignment by using the Smith-Waterman algorithm.

In SeqAn, this algorithm can be used by calling the function `localAlignment`; see Listing 24 for an example.

10.1.2 Waterman-Eggert Algorithm

Sometimes also *suboptimal local alignments* between two sequences $a_1 \ldots a_m$ and $b_1 \ldots b_n$ are of interest. Waterman and Eggert (1987) modified the Smith-Waterman algorithm such that it computes *non-intersecting* local alignments between two sequences. We say that two alignments do not *intersect*, if they have no match or mismatch in common.

WATERMANEGGERT (Algorithm 20) repeatedly calls TRACE-BACKSW on different prefixes $a_1 \ldots a_i$ and $b_1 \ldots b_j$ for decreasing $M_{i,j}$. Each call computes an alignment of score $M_{i,j}$, and the algorithm stops as soon as the score falls below a certain limit. To ensure that the computed local alignments do not intersect, the algorithm modifies M and T after each call of TRACEBACKSW: Suppose that the algorithm just computed a local alignment $\mathcal{A}$

▷ SMITHWATERMAN$(a_1 \ldots a_n, b_1 \ldots b_m)$
1   $(M, T) \leftarrow$ FILLMATRIXSW$(a_1 \ldots a_n, b_1 \ldots b_m)$
2   let $M_{i,j}$ be maximal in M
3   **return** TRACEBACKSW$(a_1 \ldots a_i, b_1 \ldots b_j, T)$

▷ FILLMATRIXSW$(a_1 \ldots a_n, b_1 \ldots b_m)$
1   $M_{0,0} \leftarrow 0$ ⎫ initiali-
2   $M_{i,0} \leftarrow 0$ for $i \in \{1, \ldots, n\}$ ⎬ zation
3   $M_{0,j} \leftarrow 0$ for $j \in \{1, \ldots, m\}$ ⎭
4   **for** $i \leftarrow 1$ **to** n **do**
5   **for** $j \leftarrow 1$ **to** m **do**
6           $M_{i,j} \leftarrow \max \begin{cases} 0 & = case_{stop} \\ M_{i-1,j-1} + \alpha(a_i, b_j) & = case_{diag} \\ M_{i-1,j} + g & = case_{up} \\ M_{i,j-1} + g & = case_{left} \end{cases}$
7           $T_{i,j} \leftarrow \operatorname{argmax}_k case_k$
8   **return** (M, T)

▷ TRACEBACKSW$(a_1 \ldots a_i, b_1 \ldots b_j, T)$
1   **case** $T_{i,j} = stop$ or $i = j = 0$:   ⎫ break
2   │   **return** $\begin{bmatrix} \ \end{bmatrix}$ ⎬ condi-
                                                                                                     ⎭ tion
3   **case** $T_{i,j} = up$ or $j = 0$:   ⎫
4   │   **return** $\left[\text{TRACEBACKSW}(a_1 \ldots a_{i-1}, b_1 \ldots b_j, T) \ \middle| \ \begin{matrix} a_i \\ - \end{matrix} \right]$
5   **case** $T_{i,j} = left$ or $i = 0$:   ⎬ recur-
6   │   **return** $\left[\text{TRACEBACKSW}(a_1 \ldots a_i, b_1 \ldots b_{j-1}, T) \ \middle| \ \begin{matrix} - \\ b_j \end{matrix} \right]$ sion
7   **case** $T_{i,j} = diag$:   │
8   │   **return** $\left[\text{TRACEBACKSW}(a_1 \ldots a_{i-1}, b_1 \ldots b_{j-1}, T) \ \middle| \ \begin{matrix} a_i \\ b_j \end{matrix} \right]$ ⎭

Algorithm 19: Smith-Waterman Algorithm. $\alpha(a_i, b_j)$ is the score for aligning a_i and b_j, g is the score for a blank.

$\triangleright$ WATERMANEGGERT$(a_1 \ldots a_n, b_1 \ldots b_m, \textit{limit})$
1    $(M, T) \leftarrow$ FILLMATRIXSW$(a_1 \ldots a_n, b_1 \ldots b_m)$
2   **repeat**
3   let $M_{i,j}$ be the maximal cell in M not jet used
4   **if** $M_{i,j} < \textit{limit}$ **then break**
5        $\mathcal{A} \leftarrow$ TRACEBACKSW$(a_1 \ldots a_i, b_1 \ldots b_j, T)$
6   Report $\mathcal{A}$
7   Recompute M and T following $\mathcal{A}$

Algorithm 20: **Waterman-Eggert Algorithm.**

which aligns the two characters a_i and b_j. Then subsequent local alignments must not align a_i and b_j, so we need to recompute $M_{i,j}$ and $T_{i,j}$ such that $case_{diag}$ (see lines 6 and 7 in FILLMATRIXSW) is forbidden there. If we change $M_{i,j}$, then we possibly also need to change $M_{i+1,j}$, $M_{i,j+1}$, or $M_{i+1,j+1}$. Waterman and Eggert recalculate only the part of M that need to be updated by enumerating them from top left to bottom right.

Listing 25 shows how to compute non-intersecting suboptimal alignments in SeqAn. Each call of function `localAlignmentNext` performs one step of the Waterman-Eggert algorithm to compute the next best local alignment.

10.2   Seed Based Motif Search

Finding exact motifs is relatively easy. For example, we will show in Section 11.2 how to use index data structures to find all common q-grams between sequences in linear time. Many efficient heuristics to find high scoring but *inexact* local alignments therefore start with such small exact (or at least highly similar) motifs, so-called seeds, and extend or combine them to get larger motifs. Probably the most prominent tool of this kind is the *Basic Local Alignment Search Tool* (BLAST) (Altschul et al. 1990), which we already discussed in Section 1.2.2, but there are many other examples like FASTA (Pearson 1990) or LAGAN (Brudno et al. 2003) (see Chapter 13).

```
Align< String<Dna> > ali;
appendValue(rows(ali), "ataagcgtctcg");
appendValue(rows(ali), "tcatagagttgc");

LocalAlignmentFinder<> finder(ali);
Score<int> scoring(2,-1,-2,0);
while (localAlignment(ali, finder, scoring, 2))
{
    cout << "Score=" << getScore(finder) << endl;
    cout << ali;
}
```

Listing 25: **Waterman-Eggert Algorithm.** Applying the Waterman-Eggert algorithm in SeqAn. The algorithm computes non-overlapping local alignments with scores better than 2.

SeqAn offers the class `Seed` for storing seeds; see Section 8.6.1. In this section, we will primarily use the specialization `SimpleSeed` of this class, which is especially designed for finding good motifs between two sequences ($d = 2$). Suppose that we store a seed that corresponds to an alignment $\mathcal{A}$ between the two substrings $a_{left_0} \ldots a_{right_0}$ and $b_{left_1} \ldots b_{right_1}$, then beside the *borders* $left_0$, $right_0$, $left_1$, and $right_1$, `SimpleSeed` also knows two boundaries *lower* and *upper* for the diagonal $j - i$ of any two aligned values a_i and b_j in $\mathcal{A}$, that is $lower \leq j - i \leq upper$. The function `bandedAlignment` can be used to retrieve an alignment for a given seed. It applies a variant of the Needleman-Wunsch algorithm (see Section 8.5.1) on $a_{left_0} \ldots a_{right_0}$ and $b_{left_1} \ldots b_{right_1}$ that is *banded* by *lower* and *upper*, i.e., it only computes such values of the matrix $M_{i,j}$ for which $j - i$ lays within these boundaries.

There are two main tasks when processing seeds: extending seeds to make them longer, and chaining several seeds together. In Section 10.2.1, we will describe how to extend seeds in SeqAn. The chaining of seeds to longer motifs will be the topic of Section 10.2.2. More details about seed-based motif search in SeqAn can also be found in Kemena (2008).

10.2.1 Extending Seeds

Let $\mathcal{S}$ be a seed. Then we call another seed $\mathcal{E}$ an *extension* of $\mathcal{S}$, if for all i holds $left_i(\mathcal{E}) \leq left_i(\mathcal{S})$ and $right_i(\mathcal{S}) \leq right_i(\mathcal{E})$. A good method for extending seeds should compute an extension $\mathcal{E}$ that scores as high as possible for a given seed $\mathcal{S}$. SeqAn supports several algorithms for seed extension (see Table 22). The function `extendSeed` extends a single seed while the function `extendSeeds` extends all seeds that are stored in a container. The user can determine the directions a seed will be extended, i.e., to the *left* or to the *right* or both. We will describe in the following only the extension to the *right*; the extension to the *left* works similarly.

`MatchExtend`	A simple extension algorithm that extends seeds until the first mismatch occurs.
`UngappedXDrop`	An X-drop extension without gaps. The algorithm extends the seed until the score falls more than a given value X.
`GappedXDrop`	An X-drop extension variant of `UngappedXDrop` that also allows gaps in the extended seed.

Table 22: **Seed Extension Algorithms.**

This simple extension method (see Algorithm 21) extends the seed until the first mismatch occurs. The algorithm does not create gaps. Listing 26 shows an example.

$\triangleright$ MATCHEXTEND $(a_1 \ldots a_m, b_1 \ldots b_n, right_0, right_1)$

1   **while** $(a_{right_0 + 1} = b_{right_1 + 1})$ and
       $(right_0 < m)$ and $(right_1 < n)$ **do**
2       $right_0 \leftarrow right_0 + 1$
3       $right_1 \leftarrow right_1 + 1$

Algorithm 21: **Match Extension.**

SEED**ab**XcdXefXXX
SEED**ab**YcdefYYYY

```
String<char> a = "SEEDabXcdXefXXX";
String<char> b = "SEEDabYcdefYYYY";
Seed<> seed(0, 0, 4);                    //left=0; length=4
extendSeed(seed, a, b, 1, MatchExtend());
cout << rightPosition(seed, 0) << endl;  //output: 6
cout << rightPosition(seed, 1) << endl;  //output: 6
```

Listing 26: **Match Extension Example.** The seed SEED is extended to the
right by **ab**; then the extension stops since X and Y do not match. The
direction of the extension was selected by setting the fourth argument of
extendSeed to 1.

MATCHEXTEND has the disadvantage that a single mismatch
stops the extension immediately, so that subsequent matches are
lost. Altschul et al. (1990) therefore preferred an extension algorithm called X-*drop extension* that allows some mismatches. An
X-*drop* is a part of an alignment that scores $\leq -X$ for a certain
value $X > 0$, where X is called the *depth* of the X-drop. The
X-*drop extension* stops extending before the alignment ends in
an X-drop. This guarantees that all drops in the extended part
of the alignment have depth $< X$, thus the complete seed may
contain an X-drop (but no $2X$-drop), especially if it was extended
into both directions.

SeqAn supports an ungapped (Algorithm 22) and a gapped (Algorithm 23) variant of this algorithm. Listing 27 shows an example
for ungapped X-drop extension.

For the gapped variant, SeqAn implements an algorithm described by Zhang et al. (2000) that applies dynamic programming
similar to the Needleman-Wunsch algorithm (see Section 8.5.1).
GAPPEDXDROP (see Algorithm 23) computes values $M_{i,j}$ for
$i \geq right_0$ and $j \geq right_1$, where $M_{i,j}$ is the score of the optimal alignment between $a_{right_0 +1} \ldots a_i$ and $b_{right_1 +1} \ldots b_j$. The
values $M_{i,j}$ are computed in ascending order of their *antidiagonal*
$k = i + j$.

$\triangleright$ UNGAPPEDXDROP $(a_1 \ldots a_m, b_1 \ldots b_n, right_0, right_1, X)$

1    $score, best \leftarrow 0$

2    $i \leftarrow 1$

3   **while** $(right_0 + i < m)$ and $(right_1 + i < n)$ **do**

4        $score \leftarrow score + \alpha(a_{right_0 + i}, b_{right_1 + i})$

5   **if** $score \leq best - X$ **then break**

6   **if** $score > best$ **then**

7           $right_0 \leftarrow right_0 + i$ } extend

8           $right_1 \leftarrow right_1 + i$ } the

9           $best \leftarrow score$ } seed

10           $i \leftarrow 1$

11   **else**

12           $i \leftarrow i + 1$

Algorithm 22: Ungapped X-Drop Extension. The function α returns the score for aligning two values.

SEED**abXcd**XefXXX

SEED**abYcd**efYYYY

```
String<char> a = "SEEDabXcdXefXXX";
String<char> b = "SEEDabYcdefYYYY";
Seed<> seed(0, 0, 4);              //left=0; length=4
Score<> scoring(1, -1, -1);
extendSeed(seed, 2, scoring, a, b, 1, UngappedXDrop());
cout << rightPosition(seed, 0) << endl;  //output: 9
cout << rightPosition(seed, 1) << endl;  //output: 9
```

Listing 27: Ungapped X-Drop Extension Example. In this example, we set $X = 2$ (this is the second argument of `extendSeed`). The seed SEED is extended to the *right* by **abXcd** and **abYcd**.

▷ GAPPEDXDROP $(a_1 \ldots a_m, b_1 \ldots b_n, right_0, right_1, X)$

1    $k \leftarrow right_0 + right_1$

2    $best, score_k \leftarrow 0$

3    $L \leftarrow right_0$

4    $U \leftarrow right_0 + 1$ } initialization

5   **while** $k < n + m$ **do**

6      $k \leftarrow k + 1$

7      $L \leftarrow \max(L, k - n)$

8      $U \leftarrow \min(U, m)$

9   **for** $i \leftarrow L$ **to** U **do**

10        $j \leftarrow k - i$

11        $M_{i,j} \leftarrow \max \begin{cases} M_{i-1,j-1} + \alpha(a_i, b_j) \\ M_{i-1,j} + g \\ M_{i,j-1} + g \end{cases}$ compute $M_{i,j}$ on antidiagonal k

12   **if** $M_{i,j} \leq best - X$ **then**

13   └ $M_{i,j} \leftarrow -\infty$

14   **if** $M_{i,j} > best$ **then**

15          $right_0 \leftarrow i$ } extend the seed

16          $right_1 \leftarrow k - i$

17      $score_k \leftarrow \max_i(M_{i,k-i})$

18   **if** $score_k = score_{k-1} = -\infty$ **then break**

19      $best \leftarrow \max(best, score_k)$

20      $L \leftarrow \min\{i | M_{i,k-i} > -\infty \text{ or } M_{i-1,k-i} > -\infty\}$

21      $U \leftarrow \max\{i + 1 | M_{i,k-i} > -\infty \text{ or } M_{i,k-i-1} > -\infty\}$

Algorithm 23: **Gapped X-Drop Extension.** $\alpha(a_i, b_j)$ is the score for aligning a_i and b_j, g is the score for a blank. We assume $M_{i,j} = -\infty$ for all i and j until $M_{i,j}$ is set in line 11.

The highest score found so far is tracked in *best*. The algorithm limits the drop depth by setting all those $M_{i,j}$ to $-\infty$ that fall below $best - X$, which means that alignments going through $M_{i,j}$ will not be continued. Instead, we only need to compute values $M_{i,j}$ with $L \le i \le U$, where the bounds L and U are computed in lines 20 and 21 in a way that all relevant values are computed. The algorithm stops if either the diagonal $n + m$ was reached, or all values in the last two antidiagonals $k - 1$ and k were assigned to $-\infty$, since in this case that all further values would also be $-\infty$. The seed is then extended to the maximum $M_{i,j}$. Listing 28 shows how to use this algorithm in SeqAn.

SEED**abXcdXef**XXX

SEED**abYcd-ef**YYYY

```
...
extendSeed(seed, 2, scoring, a, b, 1, GappedXDrop());
cout << rightPosition(seed, 0) << endl;  //output: 12
cout << rightPosition(seed, 1) << endl;  //output: 11
```

Listing 28: **Gapped X-Drop Extension Example.** The same as in Listing 27 but with GappedXDrop.

10.2.2 Combining Seeds

In this section we will show how to combine seeds to larger seeds. We discussed in Section 8.6 how seeds are threaded to get *global seed chains* by appending one seed to another.

Remember that we can *append* a seed $\mathcal{S}_j$ to another $\mathcal{S}_k$, if $right_i(\mathcal{S}_k) \le left_i(\mathcal{S}_j)$ for all $i \in \{1, 2\}$. It may also be that $left_i(\mathcal{S}_k) \le left_i(\mathcal{S}_j) \le right_i(\mathcal{S}_k) \le right_i(\mathcal{S}_j)$ for all $i \in \{1, \ldots, d\}$, then we say that $\mathcal{S}_j$ *overlaps* with $\mathcal{S}_k$, and the two seeds can be *merged*. SeqAn supports some methods for both, appending and merging seeds. In both cases, the combination of $\mathcal{S}_k$ and $\mathcal{S}_j$ is denoted by $\mathcal{S}_k \circ \mathcal{S}_j$.

$\triangleright$ LOCALCHAINING$(\mathcal{S}_1, \ldots, \mathcal{S}_n)$
1   sort $\mathcal{S}_1, \ldots, \mathcal{S}_n$ in increasing order of $right_1(\mathcal{S}_i)$
2   $D \leftarrow \{\mathcal{S}_1\}$
3   **for** $j \leftarrow 2$ **to** n **do**
4      $\mathcal{S} \leftarrow \mathcal{S}_j$
5   **for each** $\mathcal{S}_k \in D$ within the range of $\mathcal{S}_j$ **do**
6   **if** $right_1(\mathcal{S}_k) < right_1(\mathcal{S}_j) - limit$ **then**
7   report motif $\mathcal{S}_k$
8            $D \leftarrow D \setminus \{\mathcal{S}_k\}$
9   **else if** $weight(\mathcal{S}_k \circ \mathcal{S}_j) > weight(\mathcal{S})$ **then**
10            $\mathcal{S} \leftarrow \mathcal{S}_k$
11   **if** $\mathcal{S} = \mathcal{S}_j$ **then**
12         $D \leftarrow D \cup \{\mathcal{S}_j\}$
13   **else**
14         $D \leftarrow D \setminus \{\mathcal{S}\}$
15         $D \leftarrow D \cup \{\mathcal{S} \circ \mathcal{S}_j\}$
16   report all motifs $\in D$

(lines 5–10 braced:) *find best partner $\mathcal{S} \in D$ for $\mathcal{S}_j$*

Algorithm 24: **Greedy Local Chaining Heuristic.** The algorithm combines seeds as long as this benefits the score. $\mathcal{S}_k \circ \mathcal{S}_j$ is the seed that we get by merging $\mathcal{S}_j$ and $\mathcal{S}_k$ or appending $\mathcal{S}_j$ to $\mathcal{S}_k$. The constant *limit* determines the maximal distance between to seeds that may be combined.

Combining seeds is certainly useful only if the score of the resulting motif exceeds the scores of both individual seeds, so it is sufficient to consider only neighboring seeds when we are looking for seeds to combine, because seeds that are too widely separated would hardly achieve high scores. This is advantageous compared to global chaining that requires to find a predecessor for each seed, no matter its distance. LOCALCHAINING (Algorithm 24) instead considers only seeds $\mathcal{S}_k$ that are within a certain *range* relative to a given seed $\mathcal{S}_j$. This range is defined by two constants *bandwidth* and *limit* as follows: (1) the diagonal $rightdiag(\mathcal{S}_k) = right_2(\mathcal{S}_k) - right_1(\mathcal{S}_k)$ of $\mathcal{S}_k$'s right border must be at most *bandwidth* away from the diagonal $leftdiag(\mathcal{S}_j) = left_2(\mathcal{S}_j) - left_1(\mathcal{S}_j)$ of $\mathcal{S}_j$'s left border, i.e., $|\, leftdiag(\mathcal{S}_j) - rightdiag(\mathcal{S}_k)| \leq bandwidth$, and (2) the distance between the right borders of $\mathcal{S}_j$ and $\mathcal{S}_k$ is below *limit*, i.e., $|\, left_1(\mathcal{S}_j) - right_1(\mathcal{S}_k)| \leq limit$ and $|\, left_2(\mathcal{S}_j) - right_2(\mathcal{S}_k)| \leq limit$; see Figure 34. The class `SeedSet` in SeqAn implements a suitable

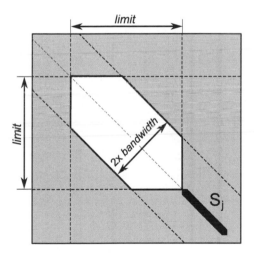

Figure 34: **Range of Possible Predecessors.** The *right end* of a predecessor $\mathcal{S}_k$ for $\mathcal{S}_j$ must be within the white area, which is: (1) *left* from $\mathcal{S}_j$, (2) between the two diagonals in distance *bandwidth* from the diagonal of the *left end* of $\mathcal{S}_j$, and (3) in distance *limit* from $left_1(\mathcal{S}_j)$.

data structure D. It stores all processed seeds in a map that allows fast searching for $\mathcal{S}_k$ that meet condition (1). Seeds that violate condition (2) are removed from D in line 6.

The application of `SeedSet` is demonstrated in Section 13.2; see the listings on page 236. One call of function `addSeed` implements the inner loop of Algorithm 24. `addSeed` offers several modes for appending or merging seeds; see Table 23. If the desired mode for adding S_j to the `SeedSet` is not possible because it contains no suitable *partner* seed S_k, then `addSeed` returns `false`.

`Single`	The seed is just added to the `SeedSet`.
`Merge`	The added seed is merged with a seed in the `SeedSet` if this benefits the score.
`SimpleChain`	The added seed is appended to a seed in the `SeedSet` if this benefits the score.
`Chaos`	The added seed is appended to a seed in the `SeedSet` if this benefits the score. Both seeds are expanded in a way that the resulting alignment contains at most one gap. The position of this gap is selected such that the score is maximized.
`Blat`	The added seed is appended to a seed in the `SeedSet` if this benefits the score. The gap between the two seeds is tried to be filled up with smaller matches.

Table 23: **Modes for Adding Seeds Using** `addSeed`.

10.3   Multiple Sequence Motifs

So far, we discussed methods for pairwise motif finding. SeqAn also implements several algorithms to find motifs in $d \geq 2$ sequences $a^1, \ldots, a^d$; see Table 24. Since the complexity for searching approximate motifs grows heavily when we want to find them in

Projection	A heuristic by Buhler and Tompa (2001) that use local sensitive hashing to get promising input estimates for the EM-algorithm; see Section 10.3.1.
EPatternBranching	This heuristic by Davila, Balla, and Rajasekaran (2006) is an extension of the *Pattern Branching* algorithm by Price et al. (2003). It applies local search techniques to optimize motif candidates. The current implementation supports only the motif models OOPS and OMOPS.
PMS1	An enumerating algorithm by Rajasekaran et al. (2005).
PMSP	A space-saving variant of PMS1 (Davila et al. 2006); see Section 10.3.2.

Table 24: **Motif Finding Algorithms.** Specializations of MotifFinder.

more than two sequences, we simplify the problem as follows:[1] (1) the length l of the wanted motif is given in advance, and (2) we allow a certain number of *mismatches* between a motif and its occurrences, but no inserts or deletes. An *occurrence* of $m = m_1 \ldots m_l$ in a string a^i is therefore a substring $a^i_{pos+1} \ldots a^i_{pos+l}$ that differs from m only in $\leq k$ values, and the occurrence is *exact*, if they differ in *exactly* k values. Figure 35 shows an example for $l = 12$ and $k = 2$.

```
0     5    10    15    20    25    30    35    40    45
TCTCATCCGGTGGGAATCACTGCCGCATTTGGAGCATAAACAATGGGGGG
TACGAAGGACAAACACTTTAGAGGTAATGGAAACACAACCGGCGCATAAA
ATACAAACGAAAGCGAGAAGCTCGCAGAAGCATGGGAGTGTAAATAAGTG
GGCGCCTCATTCTCGGTTTATAAGCCAAAACCTTGTCGAGGCAACTGTCA
TCAAATGATGCTAGCCGTCGGAATCTGGCGAGTGCATAAAAAGAGTCAAC
```

Figure 35: **Motifs Example.** Each sequence contains the motif "GGTGTATAAA" with 2 mismatches.

---

[1]Nevertheless, the problem stays NP-hard (Lanctot et al. 1999).

SeqAn offers several alternatives to define the concept *motif*, which differ in the number of required occurrences in the sequences; see Table 25. We call this the *motif model*, and together with the decision of whether exact or non-exact occurrences are to be counted, the motif model defines the kind of search. The three models OOPS, ZOOPS, and TCM were introduced by Bailey and Elkan (1995), where the latter two depend on a parameter $0 < \xi \leq 1$ that is the minimum fraction of the d sequences that must contain occurrences of m. Moreover, we implemented a new variant OMOPS (Lim 2007) that resembles TCM with $\xi = 1$.

OOPS   *one occurrence per sequence*: m is an OOPS motif, if each sequence $a^1, \ldots, a^k$ contains exactly one occurrence of m.

OMOPS   *one or more occurrence per sequence*: m is an OMOPS motif, if each sequence $a^1, \ldots, a^k$ contains at least one occurrence of m.

ZOOPS   *zero or one occurrence per sequence*: m is a ZOOPS motif, if at least ξk out of k sequences $a^1, \ldots, a^k$ contain one occurrence of m $(0 < \xi \leq 1)$.

TCM   *two-component mixture*: m is a TCM motif, if at least ξk out of k sequences $a^1, \ldots, a^k$ contain at least one occurrence of m $(0 < \xi \leq 1)$.

---

occurrences	no	one	more
OOPS	0	k	0
OMOPS	0	k	
ZOOPS	$\leq (1-\xi)k$	$\geq \xi k$	0
TCM	$\leq (1-\xi)k$	$\geq \xi k$	

Table 25: **Motif Models.** The table shows the number of sequences $a^1, \ldots, a^k$ with no, one, or more occurrence of motif m.

Motif searching in SeqAn can be accessed via `findMotif`, which get three arguments: (1) an instance of the class `MotifFinder` that specifies the applied algorithm (Table 24) stores all needed temporary data for processing the search together with all necessary constants, like the number of allowed errors k, (2) the sequences $a^1, \ldots, a^d$, and (3) a tag that specifies the motif model (Table 25). See Listing 29 for an example.

SeqAn supports two kinds of motif finding algorithms: randomized heuristics (`Projection`, `EPatternBranching`) and exhaustive enumeration algorithms (PMS1, PMSP). As an example for the former, we will sketch in the following `Projection` as an example, and PMSP for the latter.

10.3.1 The Randomized Heuristic `Projection`

Suppose that an unknown l-mer m was *planted* according to the current motif model with k or up to k errors at random positions into d random sequences $a^1, \ldots, a^d$, then finding m can be formulated as *maximum likelihood estimation: Find an estimate for m for which the chance of observing $a^1, \ldots, a^d$ is maximal*. Some tools like MEME (Bailey and Elkan 1994) and Projection (Buhler and Tompa 2001) apply an *expectation-maximization* algorithm to identify estimates for m. The *EM-algorithm* (Dempster et al. 1977) has the advantage that it can handle incomplete data, e.g., in our case, that we do not know the actual positions at which the motif was planted into the sequences. Let the *model parameters* $\theta_{h,c}$ be estimates of $p(\text{"}m_h = c\text{"})$, and define the *unknown variables* $Z_{i,j}$ as:

$$Z_{i,j} = \begin{cases} 1, \text{ if } m \text{ was planted into } a^i \text{ at position } j \\ 0, \text{ otherwise} \end{cases}$$

The algorithm optimizes $\theta_{h,c}$ by repeating two steps:

(1) E-step: We compute the expected values for $Z_{i,j}$ given the current estimates for $\theta_{h,c}$, i.e., we get according to Bayes' theorem:

$$\mathrm{E}(Z_{i,j}|a^i, \theta) \leftarrow \frac{p(a^i|Z_{i,j} = 1, \theta)}{\sum_k p(a^i|Z_{i,k} = 1, \theta)}$$

```
unsigned int const l = 4;   //length of motif
unsigned int const k = 1;   //number of substitutions

String<DnaString> dataset;
appendValue(dataset, DnaString("TCTCATCC...AATGGGGGG"));
appendValue(dataset, DnaString("TACGAAGC...ACCGATAAA"));
appendValue(dataset, DnaString("ATACAAAC...AGGTAAGTG"));
appendValue(dataset, DnaString("GGCGCCTC...CAACTGTCA"));
appendValue(dataset, DnaString("TCAAATGC...AGAGTCAAC"));

MotifFinder<Dna, PMSP> finder(l, k, true);
findMotif(finder, dataset, OMOPS());
cout << getMotif(finder) << endl; //output: "GGTGTATAAA"
```

Listing 29: **Motif Finding Example.** This example uses PMSP for searching
a motif in the sequences displayed in Figure 35. Note that the strings are
shortened for exposition. The motif model is OMOPS (specified by the third
argument of findMotif) with exact occurrences (specified by true in the
third argument of the finder's constructor.

(2) M-step: We use the values $Z_{i,j}$ from the last E-step to
re-estimate p("$m_h = c$") such that the likelihood of getting
$a^1, \ldots, a^d$ is maximized.

Dempster, Laird, and Rubin (1977) showed that the model variables θ of the EM-algorithm converge to a local maximum for the
likelihood of the observed data. Which local maximum is reached
depends on the estimate θ the algorithm starts.
PROJECTION (Algorithm 25) applies *locality-sensitive hashing* (Indyk and Motwani 1998) to determine promising inputs for the EM-
algorithm. A hashing function f is called *locality-sensitive*, if the
probability for collisions (i.e., $f(a) = f(b)$) between two hashed objects $a \neq b$ is higher for similar objects than for dissimilar objects.
Buhler and Tompa used gapped q-gram shapes (see Section 11.2.1)
as hash functions. The similarity between different occurrences of
the motif m is above the expected similarity between random l-
grams, hence if we apply locality-sensitive hashing to all l-grams
in $a^1, \ldots, a^d$, then occurrences of m have an above-average probability to collide with other l-grams, namely with other occurrences

```
    ▷ PROJECTION(a¹, ..., aᵈ, l, R, limit)
 1  repeat for up to R loop cycles
 2  │     f ← random locality-sensitive hash function
 3  │     bucket[x] ← ∅ for each hash value x
 4  │     for each l-mer w in a¹, ..., aᵈ do
 5  │     └   bucket[f(w)] ← bucket[f(w)] ∪ {w}
 6  │     for each x with | bucket[x]| ≥ limit do
 7  │     │   generate θ from bucket[x]
 8  │     │   m ← EM-ALGORITHM(θ)
 9  │     │   if m complies motif model then
10  │     │   │   report m
11  └     └   └   return
```

Algorithm 25: **Projection Algorithm for Motif Finding.**

of m. In reverse, there is a good chance for l-mers with many collisions to be occurrences of m, thus those l-mers are good input candidates θ for the EM-algorithm.

10.3.2 The Enumerating Algorithm PMSP

PMSP is an exhaustive motif search algorithm by Davila et al. (2006). SeqAn implements PMSP for all motif models; Algorithm 26 shows it for OMOPS and exact occurrences. OMOPS means that a motif occurs somewhere in a^1 with k errors, so PMSP enumerates all l-grams m with *Hamming distance* k to a substring $a^1_{j+1} \ldots a^1_{j+l}$ of a^1. If one of these m also occurs in all other sequences a^s for $s \in \{2, \ldots, d\}$, then m is a motif. The distance between m and an occurrence w^s in a^s is k, thus the distance between w^s and $a^1_{j+1} \ldots a^1_{j+l}$ is $\leq 2k$. To determine all l-grams in a^s that fulfills this condition, GET2KNEIGHBORS computes their distances to $a^1_{j+1} \ldots a^1_{j+l}$ (line 3). For $j > 0$, these distances are computed incrementally from the distances for $j - 1$.

$\triangleright$ PMSP$(a^1, \ldots, a^d, l, k)$

1   **for** $j \leftarrow 0$ **to** $length(a^1) - l$ **do**

2   **for** $s \leftarrow 2$ **to** d **do**

3         $\llcorner$ $N^s \leftarrow$ GET2KNEIGHBORS(a^1, a^s, l, k, j)

4   **for each** l-gram m with $\delta(m, a^1_{j+1} \ldots a^1_{j+l}) = k$ **do**

5   **if** for each $s \in \{2, \ldots, d\}$ exists $w^s \in N^s$ with
            $\delta(m, w^s) = k$ **then**

6   report motif m

$\triangleright$ GET2KNEIGHBORS(a, b, l, k, j)

1   $N \leftarrow \emptyset$

2   **for** $i \leftarrow length(b) - l$ **down to** 0 **do**

3       $D_i \leftarrow \begin{cases} \delta(a_1 \ldots a_l, b_{i+1} \ldots b_{i+l}) & \text{if } j = 0 \\ D_{i-1} - \delta(a_j, b_i) + \delta(a_{j+l}, b_{i+l}) & \text{otherwise} \end{cases}$

4   **if** $D_i \leq 2k$ **then**

5         $\llcorner$ $N \leftarrow N \cup \{b_{i+1} \ldots b_{i+l}\}$

6   **return** N

Algorithm 26: PMSP Algorithm for Motif Finding. Motif model is OMOPS; only exact occurrences are counted. $\delta(x, y)$ is the Hamming distance between two strings x and y.

Chapter 11

Indices

Indices are data structures that store processed data about a sequence or a set of sequences to facilitate searching in them. For example, indices allow fast exact pattern matching or exact motif finding. SeqAn implements several index data structures. Table 26 lists the available specializations of the class `Index`.

`Index_QGram`	A simple hashing table of all (gapped) *q*-grams of a string or string set; see Section 11.2.
`Index_ESA`	A *suffix array* (Manber and Myers 1990) that can be extended by a set of additional tables to an *enhanced suffix array* (Abouelhoda, Kurtz, and Ohlebusch 2002); see Section 11.3. The index implements iterators that allow using the data structure like a *suffix tree* (Weiner 1973); see Section 11.4.2.
`Index_Wotd`	A lazy suffix tree (Giegerich, Kurtz, and Stoye 1999). The index is *deferred*, which means that it is built up during the use.
`PizzaChili`	A wrapper for the index structures from the Pizza & Chili Corpus (Ferragina and Navarro 2008), e.g., for compressed text indices.

Table 26: **Index Data Structures. Specializations of** `Index`.

Many indices consist of several parts, we say it is a bunch of fibers. An enhanced suffix array (`Index_ESA`) for example has at least the fiber `ESA_Text`, which is the indexed text and the fiber `ESA_SA`, which contains the suffix table. More fibers like the *longest common prefix*-table (fiber `ESA_LCP`) can be created when needed. The

Metafunctions	
`Fiber`	The type of a fiber in the index (Tables 29 and 31).

Functions	
`getFiber`	Returns a fiber of the index.
`indexCreate`	Creates a fiber in the index.
`indexRequire`	On-demand creation of a fiber.
`indexSupplied`	Determines whether a fiber is present in the index.
`find`	Searches in the index (Section 11.1.2).

Table 27: **Common Functions and Metafunctions for Indices.**

types of fibers can be determined by the metafunction **Fiber**, and the function **getFiber** is used for accessing the fibers of an index.

String indices in SeqAn are in general capable of working on multiple sequences $a^1, \ldots, a^d$ at once. This is often done by building up the index for the concatenated string $a^1 \ldots a^d$, e.g., by using the *concatenator* of a **StringSet**; see Section 7.9 or the example at the end of the chapter.

Before we go into the algorithmic details in Section 11.2 we demonstrate how easily indices are created in SeqAn.

11.1 Working with Indices

11.1.1 Creating an Index

A substring index is a specialization of the generic class **Index** which expects 2 arguments (the second is optional). The first template argument is the type of the data structure the index should be built on. In the following, we denote this type by **TText**. For example, this could be **String<char>** to create a substring index on a string of characters:

```
Index< String<char> > myIndex;
```

Alternatively we could use **StringSet<String<char> >** to create an index on a set of character strings:

```
Index< StringSet<String<char> > > myIndex;
```

The second template argument of the **Index** class specifies the concrete implementation. In the following, we denote this type by **TSpec**. By default, this is **Index_ESA<>**, an enhanced suffix array. So, our first code example could also be written as

```
Index< String<char>, Index_ESA<> > myIndex;
```

After we have seen how to instantiate an index object, we need to know how to assign a sequence to the index. This can be done with the function **indexText** which returns a reference to a **TText** object stored in the index or directly with the index constructor:

```
// this ...
Index< String<char> > myIndex;
indexText(myIndex) = "tobeornottobe";

// ... could also be written as
Index< String<char> > myIndex("tobeornottobe")
```

11.1.2   Pattern Finding

To find all occurrences of a pattern in an indexed String or a **StringSet**, SeqAn provides the **Finder** class, which is also specialized for indices. The following example shows how to use the **Finder** class specialized for our index to search the pattern "be".

```
Finder< Index<String<char> > > myFinder(myIndex);
while (find(myFinder, "be"))
    cout << position(myFinder) << " ";
```

The output of the above program is 11 2. The finder object myFinder was created with a reference to myIndex. The function find searches the next occurrence of "be" and returns true if an occurrence was found and false otherwise. The position of an occurrence in the text is returned by the function position called with the Finder object. Please note that in contrast to online-search algorithms, the returned occurrence positions are not ascending. As you can see in the code example above, the pattern "be" was passed directly to the find function. This is a shortcut for indices and could be also written in a longer way:

```
Finder< Index<String<char> > > myFinder(myIndex);
Pattern< String<char> >        myPattern("be");
while (find(myFinder, myPattern))
    cout << position(myFinder) << " ";
```

To end this example we show how to use Finder class for both, an enhanced suffix array and a *q*-gram index.

```
#include <iostream>
#include <seqan/index.h>
using namespace seqan;

int main ()
{
```

The following code creates an Index of "tobeornottobe". As there is no second template parameter given to Index<..>, the default index based on an enhanced suffix array is used.

```
Index< String<char> > index_esa("tobeornottobe");
Finder< Index< String<char> > > finder_esa(index_esa);

cout << "hit at ";
while (find(finder_esa, "be"))
    cout << position(finder_esa) << " ";
cout << endl;
```

Now we explicitly create a *q*-gram index using an ungapped 2-gram and do the same. Instead of this fixed-size shape, you can use arbitrary shapes and assign them before calling `find` via `indexShape`.

```
typedef Index< String<char>,
            Index_QGram< UngappedShape<2> > >  TQGramIndex;
TQGramIndex index_2gram("tobeornottobe");
Finder< TQGramIndex > finder_2gram(index_2gram);

cout << "hit at ";
while (find(finder_2gram, "be"))
  cout << position(finder_2gram) << " ";
cout << endl;

return 0;
}
```

Below you see the result of calling the program.

```
user@computer:~/seqan$ cd demos
user@computer:~/scqan/demos$ make index_find
user@computer:~/seqan/demos$ ./index_find
hit at 11 2
hit at 2 11
user@computer:~/seqan/demos$
```

Now we will have a more detailed look at *q*-gram indices and suffix arrays.

11.2   *q*-Gram Indices

11.2.1   Shapes

A *q-gram* (in the narrow sense) is a string of length q, and *the* q-grams of a text $a = a_1 \ldots a_n$ are the $n - q + 1$ length-q substrings of this text. We also call this kind of q-gram ungapped since it consists of q *successive* values $a_{i+1} \ldots a_{i+q}$. A *gapped q-gram* on the other hand is a subsequence $a_{i+s_1} a_{i+s_2} \cdots a_{i+s_q}$ of a, where $s = \langle s_1, \ldots, s_q \rangle$ is an ordered set of positions $s_1 = 1 < s_2 < \cdots < s_q$. We call s a *shape*, and we define $weight(s) = q$ and $span(s) = s_q$. Ungapped q-grams are therefore a special case of gapped q-grams with the shape $s = \langle 1, \ldots, q \rangle$.

`SimpleShape`	An ungapped shape. The length q can be set at run time by calling the function `resize`.
`UngappedShape`	An ungapped shape of fixed length, i.e., the length is specified at compile time as a template argument constant.
`GappedShape`	A generic gapped shape that can be changed at run time. It is defined for example by calling the function `stringToShape` that translates a string of characters '1' (relevant position) and '0' (irrelevant gap position) into a shape, i.e., the string `"11100101"` would be translated into the shape $s = \langle 1, 2, 3, 6, 8 \rangle$.
`HardwiredShape`	This subspecialization of `GappedShape` stores a gapped shape that is defined at compile time. The shape $\langle s_1, s_2, \ldots, s_q \rangle$ is encoded in a list of the $q - 1$ differences $s_2 - s_1, s_3 - s_2, \ldots, s_q - s_{q-1}$, which are specified as template argument constants of the tag class `HardwiredShape`. For example, the shape $s = \langle 1, 2, 3, 6, 8 \rangle$ would be encoded as `HardwiredShape<1, 1, 3, 2>`.

Table 28: `Shape` **Specializations.**

SeqAn offers several alternative data structures for storing gapped or ungapped shapes; see Table 28 and Figure 11 on page 60. The main purpose of these shape classes is to compute *hash values*: Given a shape $s = \langle s_1, \ldots, s_q \rangle$, we define the hash value of a q-

gram $a_{s_1} \ldots a_{s_q}$ to be[1]

$$hash(a_{s_1} \ldots a_{s_q}) = \sum_{i=1}^{q} ord(a_{s_i})|\Sigma|^{q-i}$$

In other words, *hash* regards $ord(a_{s_1}) \ldots ord(a_{s_q})$ as a number of base $|\Sigma|$ with q digits. Obviously $hash(a^1) \neq hash(a^2)$ for two different q-grams $a^1 \neq a^2$.

The specializations of Shape differ in the performance of computing hash values; see Figure 12 on page 61. For example, the *ungapped* specializations are faster than their *gapped* counterparts if we need to compute the hash values of all q-grams in a text, since the hash value of the i-th ungapped q-gram can be incrementally computed in constant time from the hash value of the $i-1$-th ungapped q-gram by:

$$hash(a_{i+1} \ldots a_{i+q}) = hash(a_i \ldots a_{i+q-1})q - a_i|\Sigma|^q + a_{i+q}$$

Moreover, *fixed* variants are faster than their *variable* counterparts because the compiler can optimize the code better if the shape is known at compile time.

11.2.2   *q*-Gram Index Construction

Let $s = \langle s_1, \ldots, s_q \rangle$ be a shape. A q-gram index allows to look up in constant time all occurrences of a given q-gram b in a text $a = a_1 \ldots a_n$. The index consists of two tables (see Figure 36): (1) The *position table* P that enumerates the starting positions $j \in \{0, \ldots, n - s_q\}$ of all q-grams in the text ordered by $hash_s(a_{j+s_1} \ldots a_{j+s_q})$, and (2) the *directory table* D that stores for each value $k \in \{0, \ldots, |\Sigma|^q\}$ the number of q-grams b in the text with $hash_s(b) < k$. With these two tables, it is easy to look up the occurrences of b in a at the positions $P[D[hash_s(b)]], \ldots, P[D[hash_s(b) + 1] - 1]$.

Both tables can be built up together in time $O(n)$ using *count sort* (Algorithm 27). COUNTSORT sorts the positions $p_1, \ldots, p_m$ by the

---

[1]Remember that *ord* returns for each value $\in \Sigma$ a unique integer $\in \{0, \ldots, |\Sigma| - 1\}$; see Section 6.4.

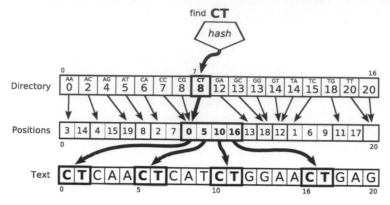

Figure 36: *q*-**gram Index.** In this example, the 2-gram index of "CTCAACTCATCTGGAACTGAG" is searched for all occurrences of "CT". We first compute $hash($"CT"$) = 7$, and then we find in the *Positions* table at $P[D[7]], \ldots, P[D[8]-1]$ the positions 0, 5, 10, and 16. (D is the *Directory* table).

keys $k_1, \ldots, k_m$ in three steps: (1) For each $k \in \{0, \ldots, Z-1\}$ it counts in D the number of $k_j = k$, (2) the counts in D are summed up such that each $D[k]$ contains the number of $k_j < k - 1$, and (3) the p_j are sorted into P guided by D. The sorting is *stable*: If $k_i = k_j$ for $i < j$, then p_i is sorted before p_j into D. BUILDQGRAMINDEX calls COUNTSORT using the hash values of all q-grams as keys. This algorithm can be implemented *lightweight*, i.e., only the space of D and P is needed, by computing the hash values *on the fly* when they are needed in the steps (1) and (3).

SeqAn also supports building up P without D, which is especially useful if $|\Sigma|^q$ gets too large. The function `createQGramIndexSAOnly` applies the `sort` algorithm from the standard C++ library. Table 29 describes the different *fibers* for q-gram based indices.

11.3 Suffix Arrays

Let $a = a_1 \ldots a_n$ be a text and $b^j = a_{j+1} \ldots a_n$ the j-th suffix of a. The *suffix array* S of a is a table that stores all positions

$\triangleright$ BUILDQGRAMINDEX $(a_1 \ldots a_n, \langle s_1, \ldots, s_q \rangle)$

1	$P \leftarrow \langle 0, 1, 2 \ldots, n - s_q \rangle$	$P = $ positions
2	$h_i \leftarrow hash(a_{i+s_1} \ldots a_{i+s_q})$ for $i \in P$	
3	$K \leftarrow \langle h_0, \ldots, h_{n-s_q} \rangle$	$K = $ keys
4	**return** COUNTSORT $(P, K, |\Sigma|^q)$	

$\triangleright$ COUNTSORT $(\langle p_1, \ldots, p_m \rangle, \langle k_1, \ldots, k_m \rangle, Z)$

1	$D[j] \leftarrow 0$ for $j \in \{0, \ldots, Z\}$	count keys
2	**for** $j \leftarrow 1$ **to** m **do**	k_j
3	$\quad$ $D[k_j + 1] \leftarrow D[k_j + 1] + 1$	
4	$count \leftarrow m$	
5	**for** $j \leftarrow Z$ **down to** 1 **do**	sum up
6	$\quad count \leftarrow count - D[j]$	counters
7	$\quad D[j] \leftarrow count$	
8	**for** $j \leftarrow 1$ **to** m **do**	sort
9	$\quad P[D[k_j + 1]] \leftarrow p_j$	positions p_j
10	$\quad D[k_j + 1] \leftarrow D[k_j + 1] + 1$	into P
11	**return** $\langle P, D \rangle$	

Algorithm 27: **Count Sort Algorithm for q-Gram Index Construction.**
The shape $\{s_1, \ldots, s_q\}$ is used for indexing a text $a_1 \ldots a_n$ of alphabet Σ.
The algorithm returns both the position table P and the directory table
D. It is assumed that $n > s_q$ and $|\Sigma|^q \geq 2$.

QGram_Text	The indexed text.
QGram_SA	The position table that stores the positions P of the q-gram occurrences.
QGram_Dir	The directory table D that allows us to look up where QGram_SA stores positions of occurrences of a given q-gram.
QGram_Shape	The shape of the q-gram that specifies which q values of a string are used to compute the hash value.

Table 29: **q-gram Index Fibers.**

$j \in \{0, \ldots, n-1\}$ in the lexicographical order $\leq_{lex}$ of the b^j. Considering that, given the text a, the suffix b^j is completely determined by the position j, we simply say that S *contains the suffixes* b^j. Figure 37 shows an example. Manber and Myers introduced this data structure in 1990.

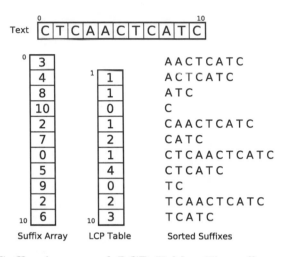

Figure 37: **Suffix Array and LCP Table.** The suffix array contains the begin positions of the lexicographically sorted suffixes of `"CTCAACTCATC"`. The LCP table stores the length of the longest common prefixes between two neighboring suffixes; see Section 11.4.1.

11.3.1 Suffix Array Construction

SeqAn implements several algorithms for constructing suffix arrays; see Table 30. Listing 30 shows an example of how to construct a suffix array in SeqAn by calling the function `createSuffixArray`.

In the following, we will discuss SKEW (see Algorithms 28, 29), a linear time algorithm by Kärkkäinen and Sanders (2003, revised in 2006), which is used in SeqAn as the default method for suffix array construction, since it is fast, generic, and is also excellent for building up suffix arrays on external memory (Dementiev et al. 2008). SKEW bases on the idea of *merge sort*: The set of suffixes

Skew3	A linear time algorithm by Kärkkäinen and Sanders (2003), which applies a merge sort approach, where two-thirds of the suffixes are recursively sorted; see SKEW, Algorithm 28, 29.
Skew7	A variant of Skew3 that recursively sorts three-seventh (instead of two-thirds) of the suffixes (Weese 2006). This reduces the number of recursive calls, so Skew7 is slightly faster than Skew3.
ManberMyers	The algorithm by Manber and Myers (1990) that is based on *prefix doubling*. It is quite slow in practice, although the run time is $O(n \log n)$.
LarssonSadakane	The algorithm by Larsson and Sadakane (2007).
SAQSort	If this tag is specified, the suffixes are sorted using the function **sort** of the standard C++ library. This is not recommended when a repetitive text is indexed.
BWTWalk	If this tag is specified, the algorithm described in Chapter 16 is used.

Table 30: **Suffix Array Construction Algorithms.**

```
String<char> text = "hello world!";
String<unsigned int> sa;
resize(sa, length(text));
createSuffixArray(sa, text, Skew7());
```

Listing 30: **Using `createSuffixArray` to Build Up a Suffix Array.** In this example, the applied algorithm is Skew7.

is divided into two parts S^{12} and S^0, each part is sorted separately (by SORTS12 and SORTS0), and then they are merged together by MERGE.

We define for $z \in \{0,1,2\}$ the sets $S^z = \{j \in S | j = z + 3i$ for integer $i\}$, and $S^{12} = S^1 \cup S^2$. SORTS12 bases on the following observation: Instead of sorting every third suffix of a, we can also sort every suffix of a string t, where each character of t consists of three successive characters of a. SORTS12 constructs values k^j for $j \in S^{12}$ that conserve the order of the first three characters of the suffixes b^j. Since we also want to consider suffixes with length < 3, we define in line 2 of Algorithm 28 the values $a_{n+1} = a_{n+2} = a_{n+3} = \$$, where $\$$ is a character not used in a for which holds $ord(\$) \leq ord(c)$ for all $c \in \Sigma$.[2] If all these k^j are different, then sorting the k^j already sorts the b^i. Otherwise (lines 6 to 15), SORTS12 applies a recursive call of SKEW to sort the suffixes of a string $t = t^1 t^2$, where t^1 and t^2 correspond to the suffixes with positions in S^1 and S^2, respectively. For minimizing the alphabet size of t, SORTS12 computes values $T(k^j)$ by enumerating the k^j, such that the $T(k^j)$ conserve the order of the k^j (lines 6 to 9). Hence the alphabet size of t is bounded by the length of the input sequence n. Note that appending t^2 to t^1 does not affect the mutual order of the suffixes in t^1, since the last character in t^1 is unique in t by construction. After that (in lines 13 to 15), the values in S^{12} are translated from positions in t into positions in a.

SORT0 (Algorithm 29) sort the suffixes of a at positions in S^0. Obviously $b^{j_1} \leq_{lex} b^{j_2}$ for two positions $j_1, j_2 \in S^0$, if either $a_{j_1+1} < a_{j_2+1}$, or $a_{j_1+1} = a_{j_2+1}$ and $b^{j_1+1} \leq_{lex} b^{j_2+1}$. The last condition was already checked in SORTS12, since $j_1 + 1, j_2 + 1 \in S^1 \subseteq S^{12}$. Therefore SORT0 first sorts the positions $j \in S^0$ according to the occurrences of $j + 1$ in S^{12} (lines 1 to 5), and then it uses stable sorting to sort them again by a_{j+1} (line 7).

MERGE scans both S^0 and S^{12}, and in each step it appends either $x = S^0[i]$ or $y = S^{12}[j]$ to S, depending on the lexicographical

---

[2]For the implementation of SKEW in SeqAn, we modified the algorithm such that it does not require a special character $\$$ (Weese 2006).

order between b^x and b^y. If $y \in S^1$, then both $x + 1 \in S^{12}$ and $y + 1 \in S^{12}$. Hence $b^{x+1} <_{lex} b^{y+1}$ if $x + 1$ comes before $y + 1$ in S^{12}, that is if $I(x + 1) < I(y + 1)$, where I is the *inverse suffix array* of S^{12}. Therefore $b^x <_{lex} b^y$ if $a_{x+1} <_{lex} a_{y+1}$ or $a_{x+1} = a_{y+1}$ and $I(x + 1) < I(y + 1)$. If on the other hand $y \in S^2$, then $x + 2, y + 2 \in S^{12}$, hence $b^x <_{lex} b^y$ if $a_{x+1}a_{x+2} <_{lex} a_{y+1}a_{y+2}$ or $a_{x+1}a_{x+2} = a_{y+1}a_{y+2}$ and $I(x + 2) < I(y + 2)$.

$\triangleright$ SKEW $(a = a_1 \ldots a_n)$

1   **if** $n = 1$ **then return** $\langle 0 \rangle$

2   **else**

3      $S^{12} \leftarrow$ SORTS12(a)

4      $S^0 \leftarrow$ SORTS0(a, S^{12})

5   **return** MERGE(a, S^{12}, S^0)

$\triangleright$ SORTS12 (a)

1   $S^{12} \leftarrow \langle 1, 2, 4, 5, \ldots, 3i+1, 3i+2, \ldots \rangle$ positions $\leq n$

2   $k^j \leftarrow hash(a_{j+1}a_{j+2}a_{j+2})$ for $j \in S^{12}$ ⎫

3   $K \leftarrow \langle k^1, k^2, k^4, k^5 \ldots, k^{3i+1}, k^{3i+2}, \ldots \rangle$ ⎬ sort 3-grams

4   $\langle S^{12}, D \rangle \leftarrow$ COUNTSORT$(S^{12}, K, |\Sigma|^3)$ ⎭

5   **if** $k^i = k^j$ for any two $i \neq j$ **then**

6      $k \leftarrow 0$ ⎫

7   **for** $j \leftarrow 0$ **to** $|\Sigma|^3 - 1$ **do**   ⎪ reduce

8        $T[j] \leftarrow k$ ⎬ alphabet

9   **if** $D[j] \neq D[j + 1]$ **then** $k \leftarrow k + 1$ ⎪ size ⎭

10      $t^1 \leftarrow T[k^1]\ T[k^4]\ T[k^7] \ldots T[t^{3i+1}] \ldots$ ⎫

11      $t^2 \leftarrow T[k^2]\ T[k^5]\ T[k^8] \ldots T[t^{3i+2}] \ldots$ ⎬ recursion

12      $S^{12} \leftarrow$ SKEW(t^1t^2) ⎭

13   **for** $j \leftarrow 0$ **to** $|S^{12}| - 1$ **do**   ⎫

14   **if** $S^{12}[j] < |t^1|$ **then** $S^{12}[j] \leftarrow 3S^{12}[j] + 1$ ⎬ transform

15   **else** $S^{12}[j] \leftarrow 3(S^{12}[j] - |t^1|) + 2$ ⎭ positions

16   remove from S^{12} all entries $\geq n$

17   **return** S^{12}

Algorithm 28: **Skew Algorithm for Suffix Array Construction (part one).** In line 2, we define $a_{n+1} = a_{n+2} = a_{n+3} = \$$. SORTS0 and MERGE are defined in Algorithm 29, COUNTSORT in Algorithm 27.

11.3.2   Searching in Suffix Arrays

Let S be the suffix array of a text $t = t_1 \ldots t_n$, $p = p_1 \ldots p_m$ a pattern, and $s^j = t_{S[j]+1} \ldots t_n$ the suffix in S at $j \in \{0, \ldots, n-1\}$. We define

$$L = \max\{j \mid s^i <_{lex} p \text{ for all } i < j\}$$
$$R = \max\{j \mid s^i \leq_{lex} p \text{ for all } i < j\}$$

Obviously $L \leq R$, and if $L < R$, then p occurs in t at positions $S[L], S[L+1], \ldots, S[R-1]$, and only there. SEARCHSA (Algorithm 30) shows how L can be found using *binary searches*. In each pass of the main loop, the interval $[left, \ldots, right]$ is cut into halves at the position mid. If $p \leq_{lex} s^{mid}$, then $L \leq mid$ and therefore $L \in [left, \ldots, mid]$, otherwise $L \in [mid+1, \ldots, right]$. The algorithm stops if the interval only contains $L = left = right$. For computing R, we need to change the condition at line 11 to

$$i \leq n \text{ and } (j > m \text{ or } p_j <_{lex} t_i),$$

which is equivalent to $p <_{lex} s^{mid}$. If this condition is fulfilled, then $R \in [left, \ldots, mid]$, otherwise $R \in [mid+1, \ldots, right]$.

SEARCHSA computes two values l and r for which the following invariants hold: (1) There is an $x \leq left$ such that p and s^x share the first $l-1$ values, and (2) p and s^{right} share the first $r-1$ values. Therefore each suffix s^{mid} with $x \leq mid \leq right$ share at least the first $\min(l, r) - 1$ values with p, so these values need no further examination when we want to compare p and s^{mid} (see lines 6 to 10). This speeds up the search, although the worst case runtime is still $\Omega(m \log n)$.

11.4   Enhanced Suffix Arrays

The suffix array can be extended to the very powerful data structure enhanced suffix array by adding further tables; see Table 31. We will discuss the LCP table (`ESA_LCP`) in Section 11.4.1. Together with the suffix table, the LCP table allows a depth-first

```
   ▷ SORTS0 (a, S^12)
1    m ← 0
2    for j ← 0 to |S^12| − 1 do
3       if S^12[j] = 3i + 1 for an integer i then
4          S^0[m] ← 3i
5          m ← m + 1
6    k^j ← ord(a_{S^0[j]+1}) for j ∈ {0,...,m − 1}
7    ⟨S^0, D⟩ ← COUNTSORT (S^0, ⟨k^0,...,k^{m−1}⟩, |Σ|)
8    return S^0
```

sort S^0 by S^1

stable resort S^0 by a

```
   ▷ MERGE (a, S^12, S^0)
1    for j ← 0 to |S^12| − 1 do I[S^12[j]] ← i
2    i, j ← 0
3    while i + j < n do
4       if i ≥ |S^0| then select ← false
5       else if j ≥ |S^12| then select ← true
6       else select ← SELECT (S^0[i], S^12[j], a, I)
7       if select then
8          S[i + j] ← S^0[i]
9          i ← i + 1
10      else
11         S[i + j] ← S^12[j]
12         j ← j + 1
13   return S
```

I = inverse S^{12}

compare $S^0[i]$ and $S^{12}[j]$

append $S^0[i]$ to S

append $S^{12}[j]$ to S

```
   ▷ SELECT (x, y, a, I)
1    if y = 3i + 1 for an integer i then
2       return a_{x+1}I[x + 1] <_{lex} a_{y+1}I[y + 1]
3    else
4       return a_{x+1}a_{x+2}I[x+2] <_{lex} a_{y+1}a_{y+2}I[y+2]
```

Algorithm 29: **Skew Algorithm for Suffix Array Construction (part two).** COUNTSORT is defined in Algorithm 27.

$\triangleright$ SEARCHSA $(p = p_1 \ldots p_m, t = t_1 \ldots t_n, S)$

1    $left \leftarrow 0$

2    $right \leftarrow n$

3    $l, r \leftarrow 1$

4   **while** $left < right$ **do**

5       $mid \leftarrow \left\lfloor \frac{left + right}{2} \right\rfloor$

6       $j \leftarrow \min(l, r)$ } compare

7       $i \leftarrow S[mid] + j$ p and s^{mid}

8   **while** $j \leq m$ and $i \leq n$ and $p_j = t_i$ **do**

9          $j \leftarrow j + 1$

10        $i \leftarrow i + 1$

11   **if** $(j > m)$ or $(i \leq n$ and $p_j <_{lex} t_i)$ **then**   } $p \leq_{lex} s^{mid}$:

12          $right \leftarrow mid$ go left

13          $r \leftarrow j$

14   **else**   } $p >_{lex} s^{mid}$:

15          $left \leftarrow mid + 1$ go right

16          $l \leftarrow j$

17   **return** $left$

Algorithm 30: **Searching a Suffix Array.** S is the suffix array of the text t, and p the pattern that is searched in t. The algorithm returns L; it would compute R if the condition in line 11 is changed to $i \leq n$ *and* $(j > m$ *or* $p_j <_{lex} t_i)$. The pattern p occurs in t at positions $S[L], S[L+1], \ldots, S[R-1]$.

search in the *suffix tree* of the text, and we will describe applications for it in Section 11.4.2.

ESA_Text	The indexed text.
ESA_SA	The *suffix array* that contains the positions of the lexicographically ordered suffixes of the indexed text ESA_Text; see Section 11.3.
ESA_LCP	A table that contains the lengths of the *longest common substrings* between adjacent suffixes in the suffix array ESA_SA; see Section 11.4.1.
ESA_ChildTab	A table that contains all structural information about the *suffix tree* that is missing in ESA_SA and ESA_LCP (Abouelhoda et al. 2002).
ESA_BWT	The Burrows-Wheeler transformation of the indexed text ESA_Text (Burrows and Wheeler 1994). It contains the preceding character a_{j-1} to each suffix $a_j \ldots a_n$ in ESA_SA.

Table 31: **Enhanced Suffix Array Fibers.**

11.4.1 LCP Table

Let S be the suffix array of the text $a = a_1 \ldots a_n$ and $b^j = a_{j+1} \ldots a_n$ the j-th suffix of a. The *LCP table* L stores in $L[k]$ the length of the *longest common prefix* between the suffix $b^{S[k]}$ and its predecessor $b^{S[k-1]}$ in S, i.e., $L[k] = lcp(b^{S[k-1]}, b^{S[k]})$ for $k \in \{1, \ldots, n-1\}$, where

$$lcp(x_1 \ldots x_k, y_1 \ldots y_l) = \max\{i \mid x_1 \ldots x_i = y_1 \ldots y_i\}.$$

The LCP table can be constructed in linear time (Kasai et al. 2001) due to the following observation: If $b^i <_{lex} b^j$ and $lcp(b^i, b^j) = h > 0$, then $b^{i+1} <_{lex} b^{j+1}$ and $lcp(b^{i+1}, b^{j+1}) = h - 1$. Any common prefix between b^{j+1} and b^{i+1} is also a prefix of the predecessor of b^{j+1} in S, thus the entry in L for b^{j+1} is $\geq h - 1$. BUILDLCPTAB (Algorithm 31) enumerates the suffixes

$b^0, \ldots, b^{n-1}$ and computes for each b^j its entry in L. The *inverse suffix array* I of S is used to determine the predecessor $b^i = b^{I[j]-1}$ of b^j in S. Suppose that the entry in L for b^i is h, then the entry for b^j in L is at least $h - 1$, so we can save $h - 1$ comparisons in line 6. Since $h \leq n - 1$, the inner loop is executed at most $2n$ times, and the runtime of BUILDLCPTAB is therefore $O(n)$.

SeqAn implements an *in-place* variant of this algorithm that does not need extra space for storing the inverse suffix array I (Weese 2006).

$\triangleright$ BUILDLCPTAB $(a_1 \ldots a_n, S)$

1	**for** $j \leftarrow 0$ **to** $n - 1$ **do** $I[S[j]] \leftarrow i$	$I =$ inverse S
2	$h \leftarrow 0$	
3	**for** $j \leftarrow 0$ **to** $n - 1$ **do**	
4	**if** $I[j] \neq 0$ **then**	
5	$i \leftarrow S[I[j] - 1]$	
6	**while** $i + h < n$ and $j + h < n$	compare
	and $a_{i+h+1} = a_{j+h+1}$ **do**	$s^{I[j]}$ and $s^{I[j]-1}$
7	$h \leftarrow h + 1$	
8	$L[I[j]] \leftarrow h$	
9	**if** $h > 0$ **then** $h \leftarrow h - 1$	
10	**return** L	

Algorithm 31: Construction of the LCP Table. S is the suffix array of $a_1 \ldots a_n$.

11.4.2 Suffix Trees Iterators

A *suffix tree* (Weiner 1973) $\mathcal{T}$ of a text $a = a_1 \ldots a_n$ is the unique rooted tree with the minimum number of vertices that have the following characteristics (see Figure 38): Let $r = v_1, v_2, \ldots, v_k = v$ be the path in $\mathcal{T}$ from the root vertex r to a vertex v, then all edges from v_{i-1} to v_i are labeled with non-empty strings $s^{i-1,i}$, and the concatenated *path label* $s^{1,2}s^{2,3} \ldots s^{k-1,k}$ is a substring of $a_1 \ldots a_n\$$, where $\$$ is a special *end-of-string* character that does not occur anywhere in a. We define $s(v)$ to be the path label of

v *without* the a trailing $ character. $\mathcal{T}$ has exactly n leaves l^i, which are labeled with the numbers $i \in \{0, \ldots, n-1\}$, such that $s(l^i) = a_{i+1} \ldots a_n$, i.e., the path labels of the leaves are the suffixes of the text. For any two leaves l^i and l^j, there is a vertex v on the paths to l^i and l^j such that $s(v)$ is the longest common prefix of $s(l^i)$ and $s(l^j)$.

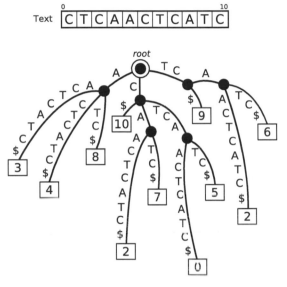

Figure 38: **Suffix Tree.** The suffix tree of "CTCAACTCATC".

The suffix tree is a versatile index data structure with many applications; see Gusfield (1997), chapters 5 to 9. SeqAn *emulates* suffix trees by the more space efficient, enhanced suffix array. An enhanced suffix array that consists of the suffix table `ESA_SA`, the LCP table `ESA_LCP`, and the *child table* `ESA_ChildTab` is capable of replacing the suffix tree in all of its applications (Abouelhoda et al. 2004). Algorithm 32 demonstrates that we only need suffix table S and LCP table L to emulate a *bottom-up traversal* of a suffix tree $\mathcal{T}$, that is, to enumerate all vertices in $\mathcal{T}$ in a way that the children appear earlier than their parents. BOTTOMUPTRAVERSAL reports for each vertex v in $\mathcal{T}$ the set of its leaves. From the construction of $\mathcal{T}$ follows that the path labels of these leaves are the suffixes that share the common prefix $s(v)$, hence the labels of the leaves are listed consecutively in the suffix array, e.g., in $S[l], \ldots, S[r-1]$, and all entries $L[l+1], \ldots, L[r-1] \geq |s(v)|$, whereas $L[l], L[r] < |s(v)|$. BOTTOMUPTRAVERSAL therefore scans L for consecutive runs of values $\geq m \in L$, and this way, it finds all inner vertices of T.

	▷ BOTTOMUPTRAVERSAL (S, L)
1	**for each** $\langle l, r, m \rangle$ reported by TRAVERSE $(S, L, 0, 0)$ **do**
2	∟ report vertex in T with leaves $S[l], \ldots, S[r-1]$

	▷ TRAVERSE (S, L, l, m)
1	$r \leftarrow l + 1$
2	**while** $r < n$ **do**
3	**if** $L[r] < m$ **then break**
4	**else if** $L[r] > m$ **then**
5	$r \leftarrow$ TRAVERSE $(S, L, r, L[r])$
6	**else** $r \leftarrow r + 1$
7	report $\langle l, r, m \rangle$
8	**return** r

Algorithm 32: Emulated Bottom-up Traversal of a Suffix Tree. S is the suffix array and L the LCP table of a length-n text. A vertex of the suffix tree is represented by the set of its leaves. The last reported vertex is the root that covers all leaves of the tree.

SeqAn supports several iterators that emulate a traversal of a suffix tree; see Table 32. We will explain how these iterators work for the example of the `SuperMaxRepeats` iterator specialization that computes all *supermaximal repeats* of the indexed text a; see Listing 31 for a code example. A *repeat* is a substring that occurs at least twice in the text, and it is *supermaximal*, if none of its occurrences is a substring of any other repeat. Let a be a text and T the suffix tree of a, then there exists a vertex v in T, such that $s(v) = w$ for each supermaximal repeat w in a. Now let v be any inner vertex of the suffix tree T of a, and let $I \subseteq \{0, \ldots, n-1\}$ be the set of occurrences of $s(v)$ in a, that is $s(v) = a_{i+1} \ldots a_{i+|s(v)|}$ for all $i \in I$. Then $|I| \geq 2$, and if for all $i, j \in I$ holds (1) $a_i \neq a_j$,

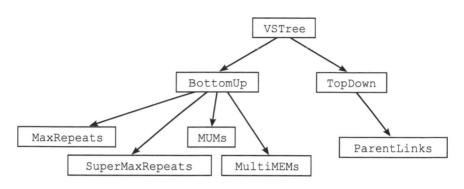

BottomUp	Generic bottom-up iterator. It enumerates the vertices of the *emulated* suffix tree during a post-order depth-first-search; see Algorithm 32.
MaxRepeats	A bottom-up iterator that enumerates all pairs of repeat occurrences that cannot be extended to the left or to the right.
SuperMaxRepeats	A bottom-up iterator that enumerates all supermaximal repeats; see Algorithm 33.
MUMs	A bottom-up iterator that enumerates maximal unique matches (MUMs) between two texts, i.e., all substrings that occur exactly once in both texts and that cannot be extended to the left or right.
MultiMEMs	Like MUMs, but for more than two texts.
TopDown	An iterator that allows to go further to any child of the current vertex. For this iterator, the child table ESA_ChildTab is required.
ParentLinks	Like TopDown but with the additional option to move from the current vertex to its parent. The iterator allows therefore any walk up and down through the *emulated* suffix tree. It requires the child table ESA_ChildTab.

Table 32: **Hierarchy of Suffix Tree Iterators.**

and (2) $a_{i+|s(v)|+1} \neq a_{j+|s(v)|+1}$, then $s(v)$ is a supermaximal repeat. This suggests the simple Algorithm 33 for finding all supermaximal repeats in a (Abouelhoda et al. 2002). SUPERMAXIMALREPEATS enumerates all inner vertices v in $\mathcal{T}$ and reports each v that fulfills both conditions (1) and (2). Note that we can check these conditions in $O(|\Sigma|)$.

$\triangleright$ SUPERMAXIMALREPEATS $(a_1 \ldots a_n, S, L)$

1   **for each** $\langle l, r, m \rangle$ reported by TRAVERSE $(S, L, 0, 0)$ **do**

2   **if** $r - l \geq 2$ **then**

3   **if** for each $i, j \in \{l, \ldots, r-1\}$ holds:
             (1) $a_{S[i]} \neq a_{S[j]}$ and
             (2) $a_{S[i]+m+1} \neq a_{S[j]+m+1}$ **then**

4   report supermaximal repeat $a_{S[l]+1} \ldots a_{S[l]+m}$

Algorithm 33: **Finding All Supermaximal Repeats in a Text.** S is the suffix array and L the LCP table of the length-n text a. In line 3, we define $a_0 \neq c$ and $a_{n+1} \neq c$ for each value $c \in \Sigma$.

The code in Listing 31 shows how the iterator is used in SeqAn. The suffix tree iterators in Table 32 differ in the way the tree nodes are traversed. After having explained the iterator for supermaximal repeats in depth, we give some examples of how to use the other iterators. For a lot of sequence algorithms it is necessary to do a full depth-first search (dfs) over all suffix tree nodes beginning either in the root (preorder dfs) or in a leaf node (postorder dfs). A preorder traversal halts in a node when visiting it for the first time, whereas a postorder traversal halts when visiting a node for the last time.

A postorder traversal, also known as bottom-up traversal, can be realized with the `BottomUp` iterator. This iterator starts in the leftmost (lexicographically smallest) leaf and provides the functions `goNext`, `atEnd`, and `goBegin` to proceed with the next node in postorder, to test for having been visiting all nodes, and to go back to the first node of the traversal. The `BottomUp` iterator can be obtained by the `Iterator` meta-function called with an

```
String<char> text = "How many wood would a woodchuck chuck.";
typedef Index< String<char> > TIndex;
TIndex idx(text);

Iterator<TMyIndex, SuperMaxRepeats>::Type it(idx, 3);
while (!atEnd(it))
{
    for (unsigned int i=0; i < countOccurrences(it); ++i)
    {
        cout << getOccurrences(it)[i] << ",";
    }
    cout << repLength(it) << ",";
    cout << representative(it) << endl;
    ++it;
}
```

Listing 31: **Searching Supermaximal Repeats.** This example finds all supermaximal repeats of length ≥ 3 in a text. For each supermaximal repeat, the program prints out the positions of its occurrences (getOccurrences), the length of the repeat (repLength), and the repeat string (representative).

Index type and the BottomUp specialization type. The following
example shows how our index can be traversed as a suffix tree with
myIterator in a bottom-up fashion:

```
// postorder dfs
Iterator< Index<String<char> >, BottomUp<> >::Type
                              myIterator(myIndex);
for (; !atEnd(myIterator); goNext(myIterator))
    // do something with myIterator
```

Another kind of traversing the suffix tree provides the light-weight
TopDown iterator. Starting in the root node the iterator can
goDown the left-most edge, the edge beginning with a certain character or the path of a certain string. goRight can be used to go
to the right (lexicographically larger) sibling of the current node.
These functions return a boolean value which indicates whether
the iterator could successfully be moved. To visit the children of
the root node in lexicographical ascending order, you could write:

```
Iterator< Index<String<char> >, TopDown<> >::Type
                              myIterator(myIndex);
goDown(myIterator);
while (goRight(myIterator))
    // do something with myIterator
```

To go back to the upper nodes, you can either save copies of
the TopDown iterator or use the more intricate TopDownHistory
iterator which stores the way back to the root and can
goUp. This is a specialization of the TopDown iterator and
can be instantiated with Iterator< Index<String<char> >
, TopDown<ParentLink<> > > ::Type myIterator(myIndex);.
As this iterator can randomly walk through the suffix tree it can
easily be used to do a preorder or postorder depth-first search.
Therefore this iterator also implements the functions goNext and
atEnd. The order of the dfs traversal can be specified with an
optional template argument of ParentLink<..> which can be
Preorder (default)

```
// preorder dfs
Iterator< Index<String<char> >,
   TopDown<ParentLink<Preorder> > >::Type myIterator(myIndex);
for (; !atEnd(myIterator); goNext(myIterator))
   // do something with myIterator
```

or `Postorder`. As top-down iterators starts in the root node, the
iterator must manually be moved down to the first postorder node
which is the left-most leaf.

```
// postorder dfs
Iterator< Index<String<char> >,
   TopDown<ParentLink<Postorder> > >::Type myIterator(myIndex);
while (goDown(myIterator));
for (; !atEnd(myIterator); goNext(myIterator))
   // do something with myIterator
```

Please note that a relaxed suffix tree is a suffix tree after removing
the $ characters and empty edges. For some bottom-up algorithms
it would be better not to remove empty edges and to have a one-
to-one relationship between leaves and suffices. In such cases you
can use the tags `PreorderEmptyEdges` or `PostorderEmptyEdges`
instead of `Preorder` or `Postorder` or `EmptyEdges` for the `TopDown`
iterator.

To end this chapter we give again some examples of how to access
suffix tree nodes to which iterators point.

11.4.2.1   How to access a suffix tree

All iterators are specializations of the general `VSTree Iterator`
class, hence they inherit all of its functions.   There are
various functions to access the node the iterator points
at, namely `representative`, `getOccurrence`, `getOccurrences`,
`isRightTerminal`, `isLeaf`, `parentEdgeLabel`, `parent`.   Note
that there is a difference between the functions `isLeaf` and
`isRightTerminal`. In a relaxed suffix tree a leaf is always a suffix, but not vice versa, as there can be internal nodes in which a
suffix ends. For them `isLeaf` returns false and `isRightTerminal`
returns true.

The following example demonstrates the usage of the iterator for MUMs:

```
#include <iostream>
#include <seqan/index.h>
using namespace seqan;
using namespace std;

int main ()
{
```

We begin with a StringSet that stores multiple strings.

```
StringSet< String<char> > mySet;
resize(mySet, 3);
mySet[0] = "SeqAn is a library for sequence analysis.";
mySet[1] = "String is the fundamental sequence type.";
mySet[2] = "Subsequences can be handled with the Segment.";
```

Then we create an Index of this StringSet.

```
typedef Index< StringSet<String<char> > > TMyIndex;
TMyIndex myIndex(mySet);
```

To find maximal unique matches (MUMs), we use the MUMs iterator and set the minimum MUM length to 3.

```
Iterator< TMyIndex, MUMs >::Type myMUMiterator(myIndex, 3);
String< SAValue<TMyIndex>::Type > occs;

while (!atEnd(myMUMiterator)) {
```

A multiple match can be represented by the positions it occurs at in every sequence and its length. `getOccurrences` returns an unordered sequence of pairs (`seqNo`,`seqOfs`) the match occurs at.

```
occs = getOccurrences(myMUMiterator);
```

To order them in ascending order with respect to `seqNo`, we use `orderOccurrences`.

```
    orderOccurrences(occs);
    for(unsigned i = 0; i < length(occs); ++i)
      cout << getValueI2(occs[i]) << ", ";
```

The function `repLength` returns the length of the match.

```
    cout << repLength(myMUMiterator) << "   ";
```

The match string itself can be determined with the function
`representative`.

```
    cout << "\t\"" << representative(myMUMiterator)
                   << '\"' << endl;

    ++myMUMiterator;
  }
  return 0;
}
```

11.4.3 Handling Multiple Sequences

Section 11.1.1 briefly described how an index of a set of strings
can be instantiated. Instead of creating an `Index` of a `String` you
can also create an `Index` of a `StringSet`. A character position of
this string set can be one of the following:

(1) A local position (default), i.e., `Pair (seqNo, seqOfs)` where
    `seqNo` identifies the string within the `StringSet` and `seqOfs`
    identifies the position within this string.

(2) A global position, i.e., single integer value between 0 and the
    sum of string lengths minus 1 (global position). This integer
    is the position in the gapless concatenation of all strings in
    the `StringSet` to a single string.

The meta-function `SAValue` determines which position type (local or global) will be used for internal index tables (suffix array,
q-gram array) and what type of position is returned by functions
like `getOccurrence` or `position`. `SAValue` returns a `Pair` which

is `local position` by default, but could be specialized to return an integer type (`global position`) for some applications. If you want to write algorithms for both variants, you should use the functions `posLocalize`, `posGlobalize`, `getSeqNo` and `getSeqOffset`. To search in multiple strings, the Finder example from above can be modified to:

```
// set StringSet
StringSet< String<char> > mySet;
resize(mySet, 3);
mySet[0] = "tobeornottobe";
mySet[1] = "thebeeonthecomb";
mySet[2] = "beingjohnmalkovich";

// find "be" in Index of StringSet
Index< StringSet<String<char> > > myIndex(mySet);
Finder< Index<StringSet<String<char> > > >
                            myFinder(myIndex);
while (find(myFinder, "be"))
    cout << position(myFinder) << "  ";
```

This code outputs:

```
< 0 , 11 >  < 1 , 3 >  < 2 , 0 >  < 0 , 2 >
```

As `TText` is a `StringSet`, `position(finder)` returns a `Pair` (`seqNo,seqOfs`) where `seqNo` is the number and `seqOfs` the local position of the sequence the pattern occurs at.

Chapter 12

Graphs

A *graph* $\mathcal{G}$ consists of a set V of *vertices* and a set $E \subseteq V \times V$ of *edges* between the vertices. $\mathcal{G}$ is called *undirected*, if for each edge $e = \langle v, u \rangle \in E$ also the reverse $\langle u, v \rangle \in E$, otherwise $\mathcal{G}$ is *directed*. For an edge $\langle v, u \rangle$ of a directed graph we say that it *goes from v to u*, and that the vertices v and u are adjacent.

Graphs are very common in computer science, and they have also many applications in sequence analysis, for example automata (Section 12.1) or alignment graphs (Section 12.2). Although SeqAn is not a declared graph library like the Boost Graph Library (Siek et al. 2002) or LEDA (Mehlhorn and Näher 1999), it offers a variety of graph types and algorithms. Graph data structures in SeqAn are implemented as specializations of the class `Graph` (see Table 33). Graphs can be traversed using iterators. Besides the fundamental graph data structures, SeqAn provides the most well-known graph algorithms, e.g., Dijkstra's shortest path algorithm. The fundamental graph algorithms are supplemented with some specialized alignment algorithms. All graphs can be exported in dot format for visualization.

Functions like `addVertex`, `removeVertex`, `addEdge`, or `removeEdge` can be used to add or remove vertices and edges. Each vertex and each edge in a graph is identified by a so-called descriptor. The usual descriptor type for vertices is `unsigned int`, it can be determined by calling the metafunction `VertexDescriptor`. The metafunction `EdgeDescriptor` returns the descriptor type for edges, which is usually a pointer to the data structure that holds information about the edge. The following example shows how to build up a simple graph:

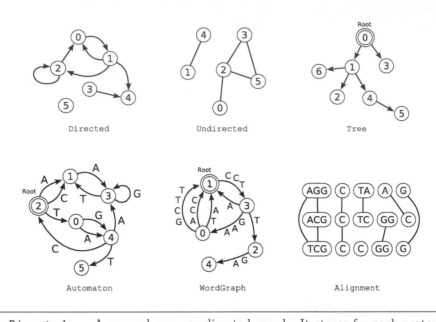

`Directed`	A general purpose directed graph. It stores for each vertex $v \in V$ an adjacency list of all vertices $u \in V$ with $\langle v, u \rangle \in E$.
`Undirected`	A general purpose undirected graph. As for `Directed`, the edges are stored in an adjacency list. Functions like `addEdge` for inserting or `removeEdge` for removing edges always affect both edges $\langle v, u \rangle$ and its reverse $\langle u, v \rangle$.
`Tree`	One vertex of this directed graph is marked as *root*. A tree can be built up *from the root to the leaves* by calling the function `addChild`.
`Automaton`	A graph with character labeled edges that can be used to scan sequences; see Section 12.1.
`WordGraph`	A sub-specialization of `Automaton` that labels the edges with sequences instead of single characters.
`Alignment`	Alignment graphs are a very flexible way of storing alignments between two or more sequences; see Section 12.2.
`Hmm`	This graph type is used to store hidden Markov models (HMMs).

Table 33: **Graph Data Structures.** Specializations of the class **Graph**.

	Metafunctions
`EdgeDescriptor`	The type that is used to store edge descriptors, i.e., handles for edges in the graph.
`VertexDescriptor`	The type that is used to store vertex descriptors, i.e., handles for vertices in the graph.
`Cargo`	The type of additional data stored in the edges of the graph, or `void` if no additional data should be stored.

	Functions
`addVertex`	Adds a new vertex to the graph.
`addEdge`	Adds a new edge to the graph
`removeVertex`	Removes a vertex from the graph.
`removeEdge`	Removes an edge from the graph.
`numVertices`	Returns the number of vertices stored in the graph.
`numEdges`	Returns the number of edges stored in the graph.
`clearVertices`	Removes all vertices from the graph.
`clearEdges`	Removes all edges from the graph.
`degree`	Returns the number of incident edges for a given vertex.
`depth_first_search`	Performs a depth-first search through the graph.
`breadth_first_search`	Performs a breadth-first search through the graph.
`resizeVertexMap`	Initializes a vertex map that can be used to store additional data attached to the vertices of the graph.
`resizeEdgeMap`	Initializes an edge map that can be used to store additional data attached to the edges of the graph.

Table 34: **Common Functions and Metafunctions for Graphs.**

```
typedef Graph< Directed<> > TGraph;
typedef VertexDescriptor<TGraph>::Type TVertexDescriptor;
typedef EdgeDescriptor<TGraph>::Type TEdgeDescriptor;

TGraph g;
TVertexDescriptor v = addVertex(g);
TVertexDescriptor u = addVertex(g);
TEdgeDescriptor e = addEdge(g, v, u);
```

The graph in the above example has data attached neither to its
vertices nor to its edges. *Property maps* are a fundamental abstraction mechanism to attach auxiliary information to the vertices and
edges of a graph. A typical example is the city names and flight
distances of a graph representing a flight network. In most sce-
narios, users should use an external property map to attach this
information to the graph, but other mechanisms do exist (below we
illustrate the use of a *Cargo* parameter). Be aware that the word
external is a hint that the information is stored independently of
the graph and functions like **removeVertex** do not affect the property map. Property maps are indexed via the already well-known
vertex and edge descriptors. This is quite handy since we can use,
for instance, a vertex iterator to traverse the graph and access
the properties on the fly. Most graph algorithms make heavy use
of this concept and therefore it is illustrated below. First, let us
create a simple graph.

```
Graph<Directed<> > g;
TVertexDescriptor v0 = addVertex(g);
TVertexDescriptor v1 = addVertex(g);
TEdgeDescriptor e1 = addEdge(g,v0,v1);
```

Second, we have to create and resize an edge property map so that
it can hold information on all edges.

```
String<unsigned int> distanceMap;
resizeEdgeMap(g, distanceMap);
```

As you can see, property maps are simply strings of some type. In
this case, we use **unsigned integer** to store the distances for our
graph g. For each edge we can store a distance now.

```
assignProperty(distanceMap, e1, 10);
```

Note how the edge descriptor is used to access the property map. An algorithm can now traverse all edges and access the distances.

```
typedef typename Iterator<Graph<Directed<> >,
                          EdgeIterator>::Type TEdgeIterator;
TEdgeIterator it(g);
for(;!atEnd(it);goNext(it)) {
    std::cout << getProperty(distanceMap, *it) << std::endl;
}
```

Alternatively a directed graph can be defined with two optional arguments, a *cargo* and a *specification*. Cargos are attached to edges in the graph. A typical example are distances for flight networks represented as a graph. The basic principle is simple: If you add an edge you have to provide a cargo, if you remove an edge you also remove the cargo. If you do not want to use cargos, you can leave out the TCargo parameter or you can use void, which is the default. All edges are directed, that is, they have a distinct source and target vertex.

```
Graph<Directed<TCargo> > directedGraph;
```

TCargo is any kind of data type, e.g., a double.

There are several *iterators* in SeqAn for traversing vertices or edges, and to traverse graphs; see Table 35. This is demonstrated by the following example program that enumerates the vertices of the graph g and prints out their descriptors:

```
typedef Iterator<TGraph, AdjacencyIterator>::Type
                            TAdjacencyIterator;
for (TAdjacencyIterator it(g); !atEnd(it) ;goNext(it))
{
    std::cout << *it << ",";
}
```

In the next example you see the use of an edge iterator. SeqAn provides a simple Edge Iterator and an Out-Edge Iterator. The use of the latter is showed in the next code piece.

```
typedef typename Iterator<TGraph, OutEdgeIterator>::Type
                                    TOutEdgeIterator;
TOutEdgeIterator it(g, v0);
for(;!atEnd(it);goNext(it)) {
    std::cout << cargo(*it) << std::endl;
}
```

In this example, we traverse all out-edges of vertex $v0$ and print their cargo.

Vertex Iterators	
VertexIterator	Enumerates all vertices of a graph in increasing order of their descriptor.
AdjacencyIterator	Enumerates for a vertex v all vertices u such that $\langle v, u \rangle \in E$.
DfsPreorder	Starting from a given vertex (e.g., the root in case of a **Tree** or **Automaton**), this iterator enumerates all reachable vertices in depth-first-search ordering.
BfsIterator	Starting from a given vertex (e.g., the root in case of a **Tree** or **Automaton**), this iterator enumerates all reachable vertices in breadth-first-search ordering.
Edge Iterators	
EdgeIterator	Enumerates all edges of a graph.
OutEdgeIterator	Enumerates for a vertex v all edges $\langle v, u \rangle \in E$.

Table 35: **Graph Iterators.** These tags are used as template arguments for the **Iterator** metafunction to select the iterator type for a **Graph** object.

SeqAn implements a variety of standard algorithms on graphs; see Table 36, most of them are described in Cormen et al. (2001). The bioinformatics algorithms heavily use the alignment graph, all others use the appropriate standard graph. All algorithms require some kind of additional input, e.g., Dijkstra's algorithm requires a distance map, alignment algorithms sequences and a score type; the network flow algorithm requires capacities on edges. The basic usage can be found in the documentation but a simple example is given here. The following code piece aligns two sequences.

```
typedef String<char> TString;
typedef StringSet<TString, Dependent<> > TStringSet;
typedef Graph<Alignment<TStringSet, void> > TGraph;

TStringSet str;
TString str0("annual");
assignValueById(str, str0);
TString str1("annealing");
assignValueById(str, str1);
TGraph g(str);
Score<int> score_type = Score<int>(0,-1,-1,0);
int score = globalAlignment(g, score_type,
                            NeedlemanWunsch() );
```

We first assign the sequences annual and annealing to a string set. Based on this string set an alignment graph is created. A Score object is used to configure the Needleman-Wunsch algorithm and the globalAlignment call creates the alignment. We will now describe in more detail the different graph specializations with a focus on algorithms for automata (Section 12.1) and alignment graphs (Section 12.2).

12.1 Automata

The specialization Automaton of Graph serves the purpose of storing *deterministic finite automata* (dfa). A *dfa* $\mathcal{G}$ is a directed graph that allows multiple edges $\langle u, v \rangle$ between two vertices u and v. The edges are labeled with characters such that two different edges $\langle u, v \rangle$ and $\langle u, v' \rangle$ going out from the same vertex u have different labels. A certain vertex called the *root* can be used as a start-

Searching	Breadth-first-search: `breadth_first_search`. Depth-first-search: `depth_first_search`. (Both can also be done by iterators; see Table 35.)
Topological Sort	`topological_sort`.
Components	`strongly_connected_components`.
Shortest Path	Single-source shortest path problem: `dag_shortest_path`, `dijkstra`, `bellman_ford_algorithm`. All-pairs shortest path problem: `floyd_warshall`.
Minimum Spanning Tree	`prims_algorithm`, `kruskals_algorithm`.
Maximum Flow	`ford_fulkerson`.
Transitive Closure	`transitive_closure`.

Table 36: **Overview of Common Graph Algorithms in SeqAn.** We omit here algorithms especially designed for automata (see Section 12.1) and alignment graphs (see Section 12.2).

ing point for a run through the automaton guided by a coincident scan through a sequence: Let $r = v_1, v_2, \ldots, v_k = v$ be a path p in $\mathcal{G}$, and let $s^{i-1,i}$ be the label of the edge from v_{i-1} to v_i, then we call $s(p) = s^{1,2}s^{2,3} \ldots s^{k-1,k}$ the *path label* of p. By definition, deterministic automata are constructed such that $s(p^1) \neq s(p^2)$ for two different paths p^1 and p^2 both starting in the root. Let $a = a_1 \ldots a_n$ be a string, then we say that a graph $\mathcal{G}$ *scans* a, if there is a path p starting from the root in $\mathcal{G}$ with $s(p) = a$. If $\mathcal{G}$ scans a, then it obviously also scans all prefixes $a_1 \ldots a_i$ of a. Finding the maximal prefix that is scanned by a given graph is part of pattern matching algorithms like BFAM (Section 9.1.4) or MULTIBFAM (Section 9.2.2), and it can also be done in SeqAn by calling the function `parseString`.

Note that `Graph` objects do not store a set of *accept states* as it is usually supposed in the literature about automata theory (e.g., Hopcroft and Ullman, 1990). If accept states are needed, then we can use a property map of `bool` to store for each vertex whether it is accepting or not.

In the following, we will discuss two special kinds of automata in more detail, namely *tries* (Section 12.1.1) and *factor oracles* (Section 12.1.2).

12.1.1 Tries

A dfa is called a trie, if it is also a tree, i.e., if the root has no incoming edges and there is for each vertex u a unique path from the root to u; see Figure 39.

The trie of a set of sequences $a^1, \ldots, a^d$ is the unique minimal trie that scans all sequences a^i. In SeqAn, the function `createTrie` implements the simple algorithm BUILDTRIE (Algorithm 34) for building up *the* trie for a given set of sequences. This takes $O(n)$ time and space, where n is the sum of the sequence lengths.

The *suffix trie* of a sequence $a = a_1 \ldots a_n$ is the trie of all suffixes $a_j \ldots a_n$. A suffix trie of a scans exactly the substrings of a. The function `createSuffixTrie` can be used to construct suffix tries.

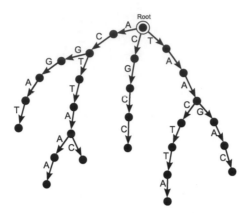

Figure 39: **Trie.** The trie of the sequences "ACGGAT", "ACTTAAA", "ACTTAC", "CGCC", "TAACTTA", and "TAAGAC".

$\triangleright$ BUILDTRIE $(a^1, a^2, \ldots, a^d)$

1 $\mathcal{G} \leftarrow$ graph that only contains the root r
2 **for** $i \leftarrow 1$ **to** d **do**
3      $n^i \leftarrow$ length of a^i
4      $k \leftarrow \max\{j \le n^i \mid \mathcal{G} \text{ scans the prefix } a_1^i \ldots a_j^i\}$
5   **if** $k < n^i$ **then**
6          $v \leftarrow$ the vertex in $\mathcal{G}$ with $s(v) = a_1^i \ldots a_k^i$
7   append to v a new branch with labels $a_k^i \ldots a_{n^i}^i$

$\left.\vphantom{\begin{array}{c}1\\2\\3\\4\\5\\6\\7\end{array}}\right\}$ add a^i to $\mathcal{G}$

Algorithm 34: **Trie Construction.** The algorithm builds up the trie of the sequences $a^1, a^2, \ldots, a^d$.

12.1.2 Factor Oracles

Suffix tries for sequences $a^1, \ldots, a^d$ of total length n have worst case size $\Omega(n^2)$. Allauzen, Crochemore, and Raffinot (1999) proposed an alternative dfa called *factor oracle* that also scans all suffixes $a^1, \ldots, a^d$, but has only $\leq n+1$ vertices and $\leq 2n$ edges. The factor oracle may – other than the *factor trie* – also scan sequences that are no substrings of any a^i. For example, the factor oracle in Figure 40 scans "CAC", which is no substring of "CTCAACTCATC".

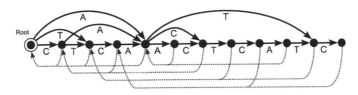

Figure 40: **Factor Oracle.** The oracle of the sequence "CTCAACTCATC". The dotted arrows visualize the supply array S.

Algorithm 35 shows how to construct a factor oracle in linear time by adding at most n more edges to the trie of $a^1, \ldots, a^d$. BUILDORACLE traverses the trie starting from the root r. Let $u \in V$ be a vertex and $s(u) = s_1 \ldots s_m$ the label of the path from r to u in the trie, let $v \in V$ be the predecessor to u on this path, and let c be the label of its last edge $\langle v, u \rangle$. When the main loop reaches u, then the algorithm extends $\mathcal{G}$ to ensure that it scans all suffixes of $s(u)$. So $\mathcal{G}$ will scan all suffixes of $a^1, \ldots, a^d$ once all vertices are processed.

Since the vertex v was processed before u, we already made $\mathcal{G}$ to scan all suffixes of $s(v) = s_1 \ldots s_{m-1}$. Let v^i be the vertex that is reached if we scan in $\mathcal{G}$ for the i-th suffix $s(v^i) = s_i \ldots s_{m-1}$, where $i \in \{0, \ldots, m-1\}$. We want $\mathcal{G}$ to scan also the suffixes $s(v^i)c$ of $s(u)$, so we just need to take care that each v^i has a c-edge, which means that v^i has an outgoing edge labeled with c. For $v^0 = v$ such an edge already exists in the trie; all other v^i can be found by following a linked list stored in the supply array S. BUILDORACLE constructs S in a way that $S[v]$ is the vertex $v^i \neq v$ with minimal i, i.e., $s(S[v])$ is the longest suffix of all suffixes $s(v^i)$

with $v^i \neq v$. All $s(v^i)$ with $v^i \neq v$ are suffixes $s(S[v])$, so the next largest suffix of $s(v)$ smaller than $s(S[v])$ is $s(S[S[v]])$. The set $\{v, S[v], S^2[v], S^3[v], \ldots, r\}$ therefore contains all vertices v^i in descending order of their suffix lengths. Enumerating the vertices u in breadth-first search order (line 4) ensures that the supply values of all vertices v^i are already computed.

For each vertex $S^j[v]$ without c-edge, we add in line 12 an edge $\langle S^j[v], u \rangle$ labeled with c. From now on $\mathcal{G}$ scans the suffix $s(S^j[v])c$ of $s(u)$. Since each scan that reaches u can go further in the trie, $\mathcal{G}$ now also scans all suffixes of $a^1, \ldots, a^d$ that start with $s(S^j[v])c$. There exists at least one suffix of that kind, and for this suffix we will never need to insert another edge into $\mathcal{G}$. The number of edges added in line 12 of BUILDORACLE is therefore bounded by the number of suffixes of $a^1, \ldots, a^d$, i.e., it is $\leq n$.

Note that we can stop the enumeration of the $S^j[v]$ as soon as we found a vertex $S^j[v]$ with a c-edge $\langle S^j[v], w \rangle$, since in this case, w is a vertex that was already processed by $\mathcal{G}$, and therefore all further vertices $S[v^i], S^2[v^i], \ldots$ have c-edges. Therefore the total runtime of BUILDORACLE is $O(n)$.

SeqAn implements this algorithm in the function `createOracle`.

12.2 Alignment Graphs

12.2.1 Alignment Graph Data Structure

Alignment graphs are, besides `Align` data structures (see Section 8.2), the second representation for alignments in SeqAn. They were initially introduced by Kececioglu (1993) and later extended by Kececioglu et al. (2000).

An *alignment graph* $\mathcal{G}$ (see Figure 41) for d sequences $a^1, a^2, \ldots, a^d$ is an undirected d-partite graph with a set V of vertices and a set E of edges that meet the following criteria:

(1) $V = V^1 \cup V^2 \cup \ldots \cup V^d$, where V^i partitions a^i into non-overlapping segments (for $1 \leq i \leq d$), i.e., each value in a^i belongs to exactly one vertex in V^i.

```
   ▷ BUILDORACLE (a¹, a², ..., aᵈ)
1    G = ⟨V, E⟩ ← BUILDTRIE(a¹, a², ..., aᵈ)
2    r ← root of G
3    S[r] ← nil
4    for each u ∈ V \ {r} in bfs order do
5        v ← the predecessor vertex of u in the trie
6        c ← label of ⟨v, u⟩
7        repeat
8            v ← S[v]
9            if exists ⟨v, w⟩ ∈ E labeled with c
             then
10               S[u] ← w
11               break
12           insert edge ⟨v, u⟩ into E with label c
13           if v = r then
14               S[u] ← r
15               break
16   return G
```

Algorithm 35: **Factor Oracle Construction.** The algorithm builds up the oracle for the sequences $a^1, a^2, \ldots, a^d$.

Alignment Matrix

a_1 - A C - - A A G - C G T A G C A

a_2 - A C - T A C G A - G - A G C A

a_3 C A C T T A T G - C C - A G - -

Alignment Graph

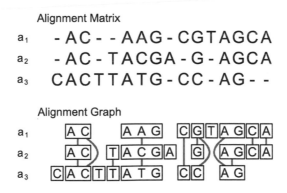

Figure 41: **Alignment Matrix and Alignment Graph.** An alignment of three sequences, displayed both in matrix style and as alignment graph.

(2) $E \subseteq \{\langle v^i, v^j \rangle \mid v_i \in V^i \text{ and } v^j \in V^j \text{ and } 1 \leq i, j \leq d \text{ and } i \neq j \text{ and the segments } v^i \text{ and } v^j \text{ have the same length}\}$.

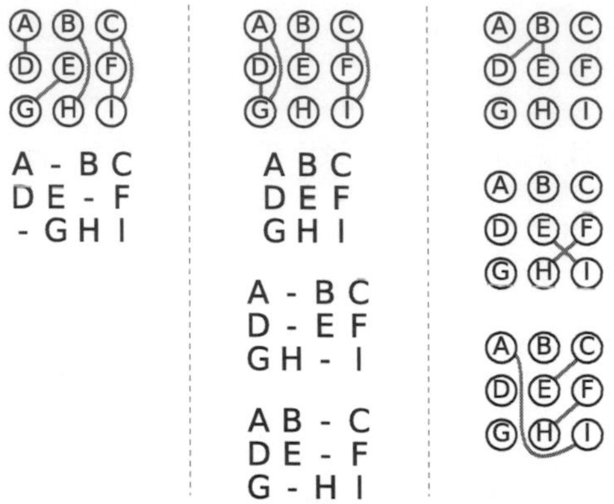

```
A - B C        A B C
D E - F        D E F
- G H I        G H I

               A - B C
               D - E F
               G H - I

               A B - C
               D E - F
               G - H I
```

Figure 42: **Alignment Graph Examples.** For the sequences "ABC", "DEF", and "GHI". **Left:** A unique trace and the compatible alignment. **Middle:** A non-unique trace and compatible alignments. The last two alignments contain changing gaps at columns 2 and 3. **Right:** Three alignment graphs that are not traces.

An alignment $\mathcal{A}$ and an alignment graph $\mathcal{G}$ are *compatible*, if for each edge $\langle v^i, v^j \rangle$ in $\mathcal{G}$, the segments v^i and v^j are aligned in $\mathcal{A}$ without gaps. An alignment graph that is compatible to at least one alignment is called a *trace*, and we call a trace unique, if it is compatible to exactly one alignment. Figure 41 and Figure 42 (left) show examples of unique traces.

Some alignments are not compatible to any unique trace, because they *contain changing gaps* (see Figure 42, middle), i.e., in the alignment are two flanking columns i and $i+1$ that together contain at most one value of each sequence. Note that optimal alignments usually do not contain changing gaps, since, for reasonable scoring schemes, the score gets better when the two columns i and $i+1$ are merged (see Section 8.3.1). For an alignment $\mathcal{A}$ that does not contain changing gaps, the alignment graph $\mathcal{G} = \langle V, E \rangle$ with $V := \{a \mid a \text{ is a value in one of the sequences of } \mathcal{A}\}$ and $E := \{\langle a, b \rangle \mid a \text{ and } b \text{ are aligned in } \mathcal{A}\}$ is a unique trace compatible to $\mathcal{A}$.

12.2.2   Maximum Weight Trace Problem

Most algorithms in SeqAn for computing optimal global alignments (Section 8.5) or local alignments (Section 10.1) accept alignment graphs for storing the results. Beside of that, we can also use alignment graphs to formulate alignment problems in a new way (Kececioglu 1993): Given an alignment graph $\mathcal{G}$ and scores $weight(e) \geq 0$ for each edge e in $\mathcal{G}$, then the *maximum weight trace problem* is to find a trace $\mathcal{G}^*$ that is a subgraph of $\mathcal{G}$ and for which the sum of the edge scores is maximal. This kind of alignment problem is especially interesting for *sparse* alignment graphs $\mathcal{G} = \langle V, E \rangle$, because in this case algorithms exist that are more efficient than, e.g., the Needleman-Wunsch algorithm (Section 8.5.1), which takes exponential time in the number of sequences.

In the following we concentrate on pairwise alignment graphs $\mathcal{G} = \langle V^1 \cup V^2, E \rangle$ between two sequences a^1 and a^2. Since the edges $\langle v^1, v^2 \rangle \in E$ can be considered as *seeds* between the two segments $v^1 \in V^1$ and $v^2 \in V^2$, and since the segments in V^1 and V^2 do not overlap, the maximum weight trace problem can be formulated as a global chaining problem; see Section 8.6.3. On the other hand, one can also reduce the maximum weight trace problem to the *heaviest common subsequence problem* (Jacobson and Vo 1992), for which SeqAn implements an efficient algorithm in function `heaviestCommonSubsequence`, that allows to compute a maximum weight trace in time $O(|E| \log |E|)$. In fact, this algorithm is equivalent to a simplified version of SPARSECHAINING (Algorithm 7, page 133), which uses *sparse dynamic programming* for global chaining; see Section 8.6.3.

V^1 is a partition of sequence a^1, and each seed covers only one segment of a^1, hence MAXWEIGHTTRACE (Algorithm 36) needs not to handle the *left* and *right* positions of the seeds separately, as it was done in SPARSECHAINING. Let for both $i \in \{1, 2\}$ the segments in the sets $V^i = \langle v_1^i, v_2^i, \ldots \rangle$ be ordered according to their occurrences in a^i, and for each edge $e_j = \langle v_p^1, v_q^2 \rangle \in E$ we define $pos_1(e_j) = p$ and $pos_2(e_j) = q$. A seed $e_j \in E$ can be *appended* to another seed $e_k \in E$, if $pos_i(e_1) < pos_i(e_2)$ for both $i \in \{1, 2\}$, and two edges $e_j, e_k \in E$ can only be part of the same trace $\mathcal{G}^*$, if one of them can be appended to the other.

$\triangleright$ MAXWEIGHTTRACE $(\mathcal{G} = \langle V, E = \{e_1, \ldots, e_n\}\rangle, w)$

1   sort $e_j \in E$ in decreasing order of $pos_2(e_j)$

2   stable sort $e_j \in E$ in increasing order of $pos_1(e_j)$

3    $D \leftarrow \emptyset$

4   **for each** $e_j \in E$ in sorted order **do**

5        $T_j \leftarrow \text{argmax}_{k \in D}\{ pos_2(e_k)|\ pos_2(e_k) < pos_2(e_j)\}$

6   **if** T_j is defined **then**

7          $M_j \leftarrow M_{T_j} + weight(e_j)$

8   **else**

9          $M_j \leftarrow weight(e_j)$ *find best chain to e_j*

10   **for each** $k \in D$ with $pos_2(e_k) \geq pos_2(e_j)$ and $M_k \leq M_j$ **do**

11          $D \leftarrow D \setminus \{k\}$ *update D*

12       $D \leftarrow D \cup \{j\}$

13    $j \leftarrow$ last element of D

14    $E^* \leftarrow \{e_j\}$

15   **while** T_j is defined **do**

16       $j \leftarrow T_j$

17       $E^* \leftarrow E^* \cup \{e_j\}$ *build new edge set E^**

18   **return** $\mathcal{G}^* = \langle V, E^* \rangle$

Algorithm 36: Maximum Weight Trace by Sparse Dynamic Programming. The algorithm is a simplified variation of Algorithm 7 on page 133. It computes a maximum weight trace subgraph $\mathcal{G}^*$ of $\mathcal{G}$, where $w(e) \geq 0$ are the weights of the edges in $\mathcal{G}$. The sorted set D stores all active seeds, M_j is the score of the best chain that ends with e_j, and T_j the predecessor of e_j in that chain. Note that argmax in line 5 returns *undefined* if it is applied to an empty set. In this case e_j has no predecessor.

The algorithm enumerates the seeds e_j in increasing order of their positions $pos_1(e_j)$ (line 4), searches in a set D of active seeds for an optimal predecessor e_{T_j} (line 5), computes the score of the best chain ending in e_j (lines 6 to 9), deletes all seeds e_k in D that are *dominated* from e_j, i.e., those e_k with $pos_2(e_k) \geq pos_2(e_j)$ and smaller chain score $M_k \leq M_j$ (lines 10 to 11), and finally inserts e_j into D (line 12). To take care that e_j is not appended to another seed e_k with $pos_1(e_k) = pos_1(e_j)$, we enumerate seeds with equal position pos_1 in decreasing order of their position pos_2 (line 1). At the end, all edges on the trace back starting from the last item in D are added to $\mathcal{G}^* = \langle V, E^* \rangle$, which is a maximal weight trace subgraph of the input graph $\mathcal{G}$.

12.2.3 Segment Match Refinement

A good strategy for getting an alignment graph $\mathcal{G} = \langle V, E \rangle$ which is *sparse enough* to be a viable input for Algorithm 36 of Section 12.2.2 is to add only those edges to V that have a good chance of becoming part of the optimal alignment, i.e., edges that connect high scoring matches between the sequences $a^1, \ldots, a^d$. If we want to construct an alignment graph $\mathcal{G}$ for a given set $S = \{\mathcal{S}_1, \ldots, \mathcal{S}_n\}$ of *seeds*, where each $\mathcal{S}_j$ aligns segments of two sequences $\in \{a^1, \ldots, a^d\}$, then this could be problematic because (1) alignment graphs allow only matches between segments of *equal length*, and (2) it is possible that two seeds $\mathcal{S}_j$ and $\mathcal{S}_k$ overlap, i.e., the aligned segments overlap. Figure 43 shows the solution

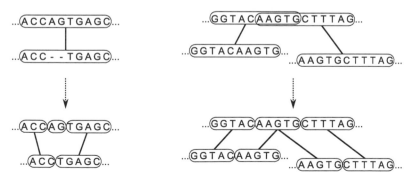

Figure 43: **Segment Match Refinement.**

for both problems: We need to cut the seeds into a set of parts $R = \{S'_1, \ldots, S'_m\}$ such that (1) each $S'_j \in R$ aligns two segments of the same length, and (2) a segment that is aligned by a seed S'_j is either identical or disjoint to the segment that is aligned by any other seed $S'_k \in R$. Condition (1) is automatically done when the alignment is transformed into its alignment graph representation. Finding a *refinement R* of minimal size that fulfills condition (2) is called the *segment match refinement problem* (Halpern, Huson, and Reinert 2002). SeqAn implements an algorithm `matchRefinement` that solves this problem for an arbitrary number of sequences (Emde 2007).

Part III

Applications

In Part III of the book we will elaborate on different aspects of programming *with* SeqAn and *in* SeqAn.

In Chapter 13 we use a well-known genome alignment program – LAGAN– to illustrate how complex analyses can be quite easily programmed in SeqAn. The original LAGAN algorithm was published in Brudno et al. (2003) and combines many algorithmic components that are now in SeqAn. In the chapter we show how LAGAN can be implemented in about 200 lines of code without losing any efficiency.

Chapter 14 is about generality. We show how versatile the multiple sequence alignment component in SeqAn is, and that it can be configured to serve a multitude of alignment tasks ranging from protein alignment to the computation of a consensus sequence in assembly projects.

The last two Chapters 15 and 16 address the algorithm engineers. In Chapter 15 the authors show how to add new functionality to SeqAn using the algorithmic components already present in the library, while in Chapter 16 the authors give a very nice example of how to incorporate a new algorithm (in this case for the construction of a suffix array) into SeqAn.

Chapter 13

Aligning Sequences with *LAGAN*

In this Chapter, we use SeqAn to re-implement the basic functionality of the common software tool LAGAN by Brudno et al. (2003).

13.1 The LAGAN Algorithm

LAGAN is a tool for aligning two long sequences $a_1 \ldots a_n$ and $b_1 \ldots b_m$, and it uses a seed chaining approach; see Section 8.6. The applied procedure (see Algorithm 37 for line numbers) works in four steps; see Figure 44:

(1) **Finding Seeds** (lines 2 to 6): For a given length $q = q_{\max}$, all common q-grams of $a_1 \ldots a_n$ and $b_1 \ldots b_m$ are found, e.g., by using a q-gram index (Section 11.2), and then combined to a set D of seeds by *local chaining* (Algorithm 24 on page 174), where the seed extension mode `Chaos` is used (Table 23 on page 176). If no common q-grams are found, the q is decreased until a minimal bound $q_{\min}$ is reached.

(2) **Global Chaining** (line 7): A *chaining* algorithm like SPARSECHAINING (Algorithm 7 on page 133) computes the optimal *global chain* $\langle \mathcal{S}_0, \ldots, \mathcal{S}_{k-1} \rangle$, where $\mathcal{S}_0$ is the *top seed* and $\mathcal{S}_{k+1}$ the *bottom seed* (see Section 8.6.2), and the rest $\mathcal{S}_1, \ldots, \mathcal{S}_k \in D$.

(3) **Filling up Gaps** (lines 8 to 12): We fill up the gaps between any two successive seeds $\mathcal{S}_i$ and $\mathcal{S}_{i+1}$ for $i \in \{0, \ldots, k\}$ by applying step (1) to (3) recursively on the gaps for a smaller

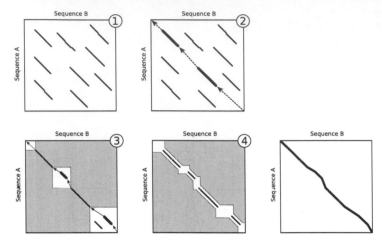

Figure 44: **The Four Steps of** LAGAN. (1) Finding seeds, (2) chaining, (3) recursively filling up gaps, and (4) banded alignment following the best chain. The result is a global alignment between the sequences.

q. This recursion stops if either the length of the gap is in both dimensions smaller than *gapsmax*, or q falls below $q_{\min}$ (line 1).

(4) **Banded Alignment** (line 2): Following the chain $\mathcal{C}$ that was computed in steps (1) to (3), a *banded alignment* algorithm (see Section 8.6.4) is used to compute a global alignment $\mathcal{A}$ between a and b.

13.2   Implementation of LAGAN

Before we start to implement Algorithm 37 in C++, we have to choose the data structures we want to use. Our objective is to align two DNA sequences a and b, so we use **String<Dna>** for storing them. The seeds $\mathcal{S}_i$ are 2-dimensional, so we apply the specialization **SimpleSeed** of **Seed** (see Section 8.6.1). These seeds are locally aligned, so the most appropriate data structure for D is **SeedSet**. We apply the *scored* variant, since this supports the

	$\triangleright$ LAGAN$(a_1 \ldots a_n, b_1 \ldots b_m)$	
1	$\mathcal{C} \leftarrow$ LAGANCHAINING(a, b, q_{max})	steps 1–3
2	$\mathcal{A} \leftarrow$ BANDEDALIGNMENT$(\mathcal{C})$	step 4
3	**return** $\mathcal{A}$	

	$\triangleright$ LAGANCHAINING$(a_1 \ldots a_n, b_1 \ldots b_m, q)$	
1	**if** $(n < gapsmax)$ and $(m < gapsmax)$ **then return** $\langle \, \rangle$	
2	$Q \leftarrow \emptyset$	
3	**while** $Q = \emptyset$ and $q < q_{min}$ **do**	
4	$\quad$ $Q \leftarrow$ all common q-grams between a and b	step 1
5	$\quad$ $q \leftarrow q - 2$	
6	$D \leftarrow$ LOCALCHAINING(Q)	
7	$\langle \mathcal{S}_0, \ldots, \mathcal{S}_{k-1} \rangle \leftarrow$ SPARSECHAINING(D)	step 2
8	**for** $i \leftarrow 0$ **to** $k+1$ **do**	
9	$\quad$ $a' \leftarrow a_{right_0(\mathcal{S}_i)} \cdots a_{left_0(\mathcal{S}_{i+1})}$	
10	$\quad$ $b' \leftarrow b_{right_1(\mathcal{S}_i)} \cdots b_{left_1(\mathcal{S}_{i+1})}$	step 3
11	$\quad$ $\mathcal{C}_i \leftarrow$ LAGANCHAINING(a', b', q)	
12	$\mathcal{C} \leftarrow \{\mathcal{S}_1, \ldots, \mathcal{S}_k\} \cup \bigcup_{i=0}^{k+1} \mathcal{C}_i$	
13	**return** sorted $\mathcal{C}$	

Algorithm 37: **The Algorithm of** LAGAN. The size q starts with q_{max} and may go down to q_{min}. LAGANCHAINING is only used if the lengths of both sequences a and b are at least *gapsmax*. For LOCALCHAINING see Algorithm 24 on page 174, and for SPARSECHAINING see Algorithm 7 on page 133. We omit the details of BANDEDALIGNMENT; see Section 8.6.4.

functionality that is used in the original tool. For storing the chains $\langle S_0, \ldots, S_{k-1} \rangle$ and C_i, we need a container class that supports fast insertion operation for merging several chains in line 12 of Algorithm 37, so a list type would be a good choice. We decide to use the class `std::list` from the standard library.

The following program consists of two functions:[1] The `main` function that implements LAGAN of Algorithm 37 and `laganChaining` that implements LAGANCHAINING.

We start the `main` function by loading the two input sequences a and b from FASTA files. For that purpose, we use `FileReader` strings (Section 7.11) that are copied to *in-memory* strings of type `String<Dna>` for speeding up the further processing:

```
typedef String<Dna> TString;
TString a = String<Dna, FileReader<Fasta> >(argv[1]);
TString b = String<Dna, FileReader<Fasta> >(argv[2]);
```

Then we call the function `laganChaining`, which is described below, to perform steps (1) to (3) of Algorithm 37:

```
typedef Seed<int, SimpleSeed> TSeed;
std::list<TSeed> chain;

laganChaining(chain,
              infix(a, 0, length(a)),
              infix(b, 0, length(b)), q_max);
```

The first argument is a list in which `laganChaining` will return a chain of seeds. Since in step (3) of the algorithm, the function will be called repeatedly on varying substrings of a and b, it expects the input sequences to be passed as *segment* objects (Section 7.7). The `main` function conveys the complete sequences.

The last argument is the length of the q-grams `laganChaining` will start with. The initial call of `laganChaining` sets the size of the q-grams to 13, and this q may fall down to `q_min`= 7 during the execution.

---

[1]This program is based on Carsten Kemena's Master's thesis (Kemena 2008).

Step (1)

In `laganChaining`, we need three data structures to perform step (1) of Algorithm 37: A seed set for storing and merging the seeds, a q-gram index for the input sequence b, and a finder for searching the q-gram index:

```
typedef typename Value<TSeed>::Type TPosition;
typedef SeedSet<TPosition, SimpleSeed, DefaultScore> TSeedSet;
TSeedSet seedset(limit, score_min, scoring_scheme);

typedef Index<TSegment, Index_QGram<SimpleShape > > TQGramIdx;
TQGramIdx index_qgram(b);

typedef Finder<TQGramIdx> TFinder;
TFinder finder(index_qgram);
```

The constants `limit` and `score_min` define the area in which local chaining searches for predecessor seeds (Section 10.2.2). The local chaining also needs `scoring_scheme` to compute scores of seeds. As long as no seeds are found, and q is at least `q_min`, we search for common q-grams in a and b and add them to `seedset`:

```
while (length(seedset) == 0)
{
    if (q < q_min) return;

    resize(indexShape(index_qgram), q);

    for (int i = 0; i < length(a)-q+1; ++i)
    {
        while (find(finder, infix(a, i, i+q)))
        {
            // add q-gram to seedset
            ...
        }
        clear(finder);
    }

    q-=2;
}
```

The variable i iterates through all q-gram positions in a, and the finder then enumerates all occurrences of the i-th q-gram of a in the indexed sequence b.

In the inner while loop, we compute the starting positions of the common q-grams relative to the complete sequences, and then add the q-gram to seedset:

```
// add q-gram to seedset
typedef typename Position<TFinder>::Type TPosition;
TPosition a_pos = beginPosition(a)+i;
TPosition b_pos = beginPosition(b)+position(finder);

if (!addSeed(seedset, a_pos, b_pos, q, 0, Merge()))
if (!addSeed(seedset, a_pos, b_pos, q, host(a), host(b),
                                      bandwidth, Chaos()))
    addSeed(seedset, a_pos, b_pos, q, Single());
```

So we first try to merge the new q-gram S with another overlapping q-gram on the same diagonal. If no such q-gram is available in seedset, then we try to find instead a predecessor S' within the area defined by limit and score_min. If a suitable S' is found, then we merge S' and S to a single seed in the *chaos style*, i.e., with a single gap in between (Table 23 on page 176). Otherwise we just add S to seedset.

Step (2)

Step (2) of Algorithm 37 just takes a single line of code:

```
globalChaining(seedset, chain);
```

The function globalChaining uses sparse dynamic programming to compute the optimal chain of seeds without a penalty for the gaps between the seeds (Section 8.6.3). Note that the resulting chain that is stored in chain does *not* contain a *top seed* or a *bottom seed*, but only the *inner seeds* from the chain.

Step (3)

Note that q was decremented at least once during step (1). If q is still $\geq$ q_min, then we enumerate all gaps between two succeeding seeds in chain and try to fill them up by recursive calls of laganChaining. After each call, all new seeds that are returned in subchain are inserted into chain between the seeds *it and seed *it2:

```
list<TSeed> subchain;
typedef typename list<TSeed>::iterator TIterator;

TIterator it = chain.begin();
TIterator it2 = it;
++it2;

while (it2 != chain.end())
{
    laganChaining(subchain,
        infix(host(a), rightDim0(*it), leftDim0(*it2)),
        infix(host(b), rightDim1(*it), leftDim1(*it2)), q);
    chain.splice(it2, subchain);

    it = it2;
    ++it2;
}
```

Note that we have to do the same for the gaps before the first seed and behind the last seed in chain.

Step (4)

Back in the main function, it remains last step (4) of Algorithm 37: We add *a* and *b* as rows to an Align object (Section 8.2) and call bandedChainAlignment (see Section 8.6.4):

```
Align<TString, ArrayGaps> alignment;
resize(rows(alignment), 2);
setSource(row(alignment, 0), a);
setSource(row(alignment, 1), b);
int score = bandedChainAlignment(chain, B, alignment,
                                 scoring_scheme);
```

The constant B is the used band width, and `scoring_scheme` defines the applied scoring scheme.

At the end, we print out the resulting alignment and its score:

```
cout << "Score: " << score << endl;
cout << alignment << endl;
```

13.3   Results

The original tool was published in 2003 by Brudno et al., and it was implemented in a combination of C programs that were stitched together by several Perl scripts. Its source code is much more extensive than our program, which takes about one hundred lines of code; for example the source code of the tool *chaos* that is responsible for steps (1) and (2) of the algorithm is more than twenty-fold larger than our program. Although the original tool is certainly more elaborate and therefore more complex than ours, both programs compute alignments of similar quality, and Figure 45 shows that the running times are also comparable.

This example shows that programs which were developed with SeqAn can match up with *hand written* tools. Moreover we demonstrated the components provided SeqAn are indeed useful for tool design, and that the application of SeqAn leads to concise and comprehensible solutions.

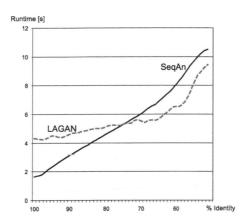

Figure 45: **Runtimes of** LAGAN **and SeqAn.** We aligned a 100kbp part of the genome of *Eschcrichia coli* with a point mutated counterpart. The figure shows the average run times (in seconds) of the original LAGAN tool and the SeqAn program depending on the similarity between the two sequences.

Chapter 14

Multiple Alignment with Segments

Anne-Katrin Emde[*] and Tobias Rausch[*]

Multiple sequence alignment is another fundamental bioinformatics task that we can easily solve within SeqAn. In the following we will see how to plug together a progressive multiple alignment strategy applicable to a large range of alignment problems.

14.1 The Algorithm

SeqAn offers all the components necessary for implementing a T-Coffee-like (Notredame, Higgins, and Heringa 2000) consistency-based multiple alignment algorithm within only 60 lines of source code. The central data structure that we make use of is the `AlignmentGraph` (Section 12.2). It allows us to compute multiple alignments on protein sequences as well as on long genomic sequences, since time and space complexity scale with the number of nodes in the graph, independent of the length of the original sequences. The algorithm is shown in Listing 38 on which we will elaborate now.

Our starting point is a set of sequences $\mathcal{S} = \{S^1, \ldots, S^n\}$. In order to compute a multiple sequence alignment, the algorithm proceeds in four steps as illustrated in Figure 46 and outlined in the following (for more details see Rausch et al. 2008):

(1) **Generating Segment Matches** (line 1): For each pair of sequences $S^p, S^q \in \mathcal{S}$ a set of local alignments, i.e., segment

---

[*]Institute for Computer Science and Mathematics, Freie Universität Berlin, Germany.

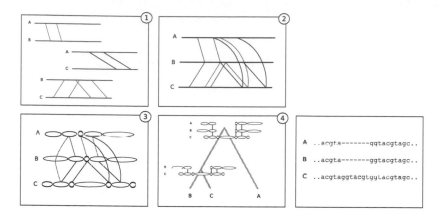

Figure 46: **Illustration of the Segment Based Multiple Alignment Method.** (1) Segment match generation, (2) segment match refinement, (3) consistency enhanced alignment graph construction, and (4) segment based progressive alignment along the guide tree. The result is a global multiple alignment of sequences A, B and C.

	$\triangleright$ SEGMENTMSA$(S^1 \dots S^n)$
1	$M \leftarrow$ LOCALMATCHES$(S^1 \dots S^n)$
2	$G \leftarrow$ MATCHREFINEMENT(M)
3	TRIPLETEXTENSION(G)
4	$D \leftarrow$ DISTANCEMATRIX(G)
5	$T \leftarrow$ GUIDETREE(D)
6	$A \leftarrow$ FOLLOWGUIDETREE(G, T)
7	**return** A

Algorithm 38: **The Algorithm of segmentMSA.** LOCALMATCHES can be any pairwise alignment method that returns the found alignments as gapless segment matches. The minimal refinement as explained in Section 12.2.3 is then computed by MATCHREFINEMENT. TRIPLETEXTENSION updates the alignment graph edges according to the consistency extension method introduced in the T-Coffee algorithm. DISTANCEMATRIX and GUIDETREE compute a hierarchical clustering as explained in Section 1.2.3 and FOLLOWGUIDETREE progressively aligns the sequences along the guide tree.

matches, is either computed or parsed from a precomputed file (for example, Blast output). Different sources of segment match information can also be combined. All segment matches are stored in a set $\mathcal{M} = \{M_1, \ldots, M_m\}$.

(2) **Segment Match Refinement** (line 2): The pairwise segment matches in $\mathcal{M}$ can overlap. We use the segment match refinement algorithm (12.2.3) to minimally subdivide them such that they can be stored in an alignment graph $G = \langle V, E \rangle$. Each node $v \in V$ corresponds to a sequence segment and each edge $e \in E$ represents part of an original input match. Each edge $e = (v_1, v_2)$ with $v_1, v_2 \in V$ is assigned a weight $w(e)$ representing the benefit of aligning the incident nodes.

(3) **Consistency Extension** (line 3): In this step, the edge weights of G are updated, such that edges (i.e., refined segment matches) consistent with other edges (i.e., other refined segment matches) are rewarded. For each triplet $v_1, v_2, v_3 \in V$ with $e_1 = (v_1, v_2), e_2 = (v_2, v_3) \in E$ and v_1 and v_3 belonging to two different sequences, the weight of edge $e_3 = (v_1, v_3)$ is increased by adding $min\{w(e_1), w(e_2)\}$. If e_3 does not yet exist, it is created with weight $min\{w(e_1), w(e_2)\}$. The result is an alignment graph $G' = \langle V, E' \rangle$ with $E \in E'$ containing a consistency-enhanced match *library* that can serve as a scoring system for the subsequent progressive multiple alignment.

(4) **Segment-Based Progressive Alignment** (line 4 to 6): Progressive alignment requires a guide tree that dictates the order in which sequences are incorporated into the growing multiple alignment (see Section 8.5.5). In order to construct such a guide tree, using the *neighbor-joining* or UPGMA algorithm, a matrix D of distances between each pair of sequences is needed. D can be filled by using the pairwise alignment information contained in G' or for example by counting the number of shared k-mers for each sequence pair. Finally, the sequences are aligned following the order

of the guide tree (8.5.5). The alignment is computed on the sequences of segments as given in G'. The alignment of two nodes v_1 and v_2, or more precisely the gapless alignment of the two subsequences stored in these nodes, is then scored with edge weight $w((v_1, v_2))$. The computation time therefore depends on the cardinality of V, and not on the total length of the original sequences. The result of this last step is an alignment graph $A\langle V, E'' \rangle$ that is 1) a subgraph of G' and 2) a valid alignment, i.e., a trace that can be converted into an alignment matrix.

14.2 Implementation

As already mentioned above, the main data structure that lends itself to addressing segment-based multiple sequence alignment in SeqAn is the `AlignmentGraph` (12.2). For storing the set of sequences, we can use a `StringSet<String<TAlphabet> >` where `TAlphabet` is the alphabet type of the sequences, e.g., `Dna` for genomic sequence alignment. First, we need to import the sequences which we suppose to be given in Fasta format.

```
typedef String<Dna> TSequence;
typedef StringSet<TSequence> TSequenceSet;

TSequenceSet seqs;
for(unsigned i = 1; i < argc; ++i)
  appendValue(seqs, String<Dna, FileReader<Fasta> >(argv[i]));
```

Step (1)

Now we create segment matches. For genomic sequence alignment, maximal unique matches (MUMs) are a reasonable choice[1]. After building an index of the sequences, we can conveniently iterate

---

[1] A suitable MUM length should be chosen according to the length and similarity of the sequences. Here, length 5 is chosen solely for illustration purposes.

over all MUMs using the appropriate `Iterator<TIndex,MUMs>`.
Function `getOccurences` returns the suffix array entries of the
individual MUM occurrences. All pairwise matches of each MUM
are then stored as `Fragment` objects in the container `matches`.
The `Fragment` data structure is a space-efficient way of storing
gapless pairwise alignments, as only the two sequence IDs and the
two start positions as well as the alignment length are stored.

```
typedef Fragment<> TMatch;
String<TMatch> matches;

typedef Index<TSequenceSet> TIndex;
TIndex index(seqs);

Iterator<TIndex,MUMs>::Type mumIt(index, 5);
String<SAValue<TIndex>::Type> occs;

while (!atEnd(mumIt))
{
  occs = getOccurrences(mumIt);
  for(unsigned i = 0; i < length(occs); ++i)
  {
    for(unsigned j = i+1; j < length(occs); ++j)
    {
      TMatch m(getValueI1(occs[i]),getValueI2(occs[i]),
               getValueI1(occs[j]), getValueI2(occs[j]),
               repLength(mumIt));
      appendValue(matches,m);
    }
  }
  ++mumIt;
}
```

Step (2)

An alignment graph is initialized on the set of sequences. As we
do not want to copy the whole sequence set, we use a dependent
`StringSet`. The alignment graph is automatically filled with the
refined segment matches by handing `matches`, `sequences` and `g` to
the function `matchRefinement`. As all match information is now
stored in g, we can discard `matches`.

```
typedef StringSet<TSequence,Dependent<> > TDepSequenceSet;
typedef Graph<Alignment<TDepSequenceSet> > TAlignmentGraph;

TAlignmentGraph g(seqs);
matchRefinement(matches, seqs, g);
clear(matches);
```

Step (3)

The TRIPLETLIBRARYEXTENSION increases the weights of consistent edges, as described in detail before.

```
tripletLibraryExtension(g);
```

Step (4)

There are many ways to compute pairwise distances. One way is to use `getDistanceMatrix` with tag `KmerDistance()` which counts the number of k-mers shared by each pair of sequences and returns a distance matrix. The guide tree of type `Graph<Tree>` can then be constructed with a hierarchical clustering method such as `upgmaTree`. Along this guide tree the sequences are aligned progressively, using g as *library* and writing the resulting trace into gOut. The result of this last step is the output alignment graph gOut that is converted to an alignment matrix when printed on screen.

```
typedef String<double> TDistanceMatrix;
TDistanceMatrix distanceMatrix;
getDistanceMatrix(g, distanceMatrix,KmerDistance());

typedef Graph<Tree<double> > TGuideTree;
TGuideTree guideTree;
upgmaTree(distanceMatrix, guideTree);

TAlignmentGraph gOut(seqs);
progressiveAlignment(g, guideTree, gOut);
cout << gOut;
```

Example

To show that this simple program produces reasonable multiple alignments for highly similar sequences, we demonstrate an exemplary run on three input sequences, seq1.fa:

```
>seq1
ACGTGGTACCCCCCGTAATAGTACAGTAT
```

seq2.fa:

```
>seq2
ACGTGGTAGTAATAGTACAGTAT
```

and seq3.fa:

```
>seq3
ACGTGGTAGTAATAGTACAGTATACAGTAT
```

Passing these three files to our demo program, we receive the following output:

```
user@computer:~$./segmentalignment seq1.fa seq2.fa seq3.fa
Alignment matrix:
      0     .     :     .     :     .     :     .
        ACGTGGTACCCCCCGTAATAGTACAGTAT-------
        ||||||||       |||||||||||||||
        ACGTGGTA------GTAATAGTACAGTAT-------
        ||||||||       |||||||||||||||
        ACGTGGTA------GTAATAGTACAGTATACAGTAT
```

14.3 Results

The above presented algorithmic pipeline has been implemented and published in two tools: SeqAn::T-Coffee (Rausch et al. 2008) and SeqCons (Rausch et al. 2009).

The first is an extension of the T-Coffee tool that not only significantly improves upon the original T-Coffee's time consumption, but is also among the fastest compared to other state-of-the-art mutliple alignment tools. On several benchmark sets such as BAliBase (Thompson, Koehl, Ripp, and Poch 2005), we could show SeqAn::T-Coffee's (and especially SeqAn::M-Coffee's; see Rausch et al. 2008) superiority in computing highly accurate protein alignments. Due to its segment-based approach, SeqAn::T-Coffee is also suitable for long genomic sequence alignment.

SeqCons is a tool for robust consensus sequence construction in genome assembly projects. Given an initial read layout, it computes multi-read alignments using the consistency- and segment-based approach. This approach has proved to be less susceptible to insertion/deletion-misalignments commonly observed in other consensus methods.

Chapter 15

Basic Statistical Indices for SeqAn

Simona E. Rombo*, Filippo Utro[†] and Raffaele Giancarlo[†]

This chapter describes in detail some functionalities included in SeqAn for the computation of fundamental *statistical indices* for strings, namely, *expectation*, *variance* and *z-score*. Their computation rests on the assumption that the strings are generated by a *Markov Chain* of finite order. The implementation of those functionalities follows closely the design principles of SeqAn and makes full use of its basic data types and algorithmic primitives, while adding new ones.

15.1 Statistical Indices and Biological Sequence Analysis

The detection of recurrent patterns in sequences is a fundamental task in data mining, in particular for biological sequences (Chen and Lonardi 2009). In this latter context, the identification of regularities in sequences cannot be separated from their biological relevance. Usually, the algorithms return a set of *interesting* relationships involving a set of substrings occurring in a given set of strings. The level of *interest* is established via statistical indices quantifying how unusual the discovered relationships are. Such an approach that has been adopted as a *de facto* paradigm in computational biology, is justified by experimental studies indicating

---

*Dipartimento di Elettronica, Informatica e Sistemistica, Università della Calabria, Italy.
†Dipartimento di Matematica ed Applicazioni, Università degli Studi di Palermo, Italy.

that there is an excellent connection between the statistical properties characterizing some substrings occurring in a set of strings and their involvement in important biological processes. Two examples may be of use in illustrating this point, while we refer the reader to (Reinert, Schbath, and Waterman 2005) for a review of the fundamental mathematical approaches and techniques characterizing the mentioned paradigm.

The well-known **BLAST** alignment program (Altschul et al. 1990) assigns a significance score to an alignment indicating how likely it is for that alignment to be due to chance. Only the ones deviating substantially from chance are returned to the user since common practice indicates that they are very likely to describe a relevant biological relation between the two strings. Chance or randomness of an alignment is established with the use of a background probabilistic model describing how likely it is that two symbols are aligned at random. In the realm of pattern discovery for the identification of transcription factors DNA binding sites (I-TFBS, for short) in herpes virus genomes, Leung et al. (Leung et al. 1996) give substantial evidence that there is a strong correlation between the number of occurrences of a substring in a string deviating from its *expected number* and that substring being a genuine binding site. Following that work, most of the very successful pattern discovery algorithms for I-TFBS are based on the identification of exceptionally frequent or rare substrings in biological sequences. That is usually done by discovering over- and under-represented patterns in biological sequences, where over- and under-representation is measured by statistical indices, e.g., *z-score*, and with the use of a background model for the data. The interested reader can find a compendium of relevant results in (Apostolico et al. 2003; Tompa et al. 2005).

For the purposes of this chapter, a particularly important incarnation of the pattern discovery process just described for I-TFBS is due to Sinha and Tompa (2003). Indeed, their method computes a *z-score* and assumes a Markovian background model of order three for the data. It is worth recalling that, although a z-score is a quite standard measure of *deviation from expectation,* its computation becomes quite challenging when applied to strings.

That is due to the computation of the *variance*, which is a far from trivial mathematical problem with only computationally costly solutions. In fact, those difficulties have led to the proliferation of algorithms for the computation of z-scores for strings, (e.g., Apostolico et al. 2003; Apostolico and Pizzi 2007), where the main differences among them are on how the variance is computed and on the basic probabilistic assumptions on how strings are generated via a background probabilistic model. The approach by Sinha and Tompa offers the advantages of being both very rigorous from the mathematical point of view and inclusive of a generalization of z-score computation to genomic sequences that accounts for their specific nature, in particular complementary base-pairing. It also lends itself to a very clean implementation in terms of classic string matching primitives.

Before we enter the technical part of this chapter, it should be pointed out that the material presented here is also strongly related to Chapter 10, dealing with the problem of motif searching in molecular biology.

15.2 Mathematical Outline

Let Σ be a finite alphabet from which strings are generated. Let W be a multi-set of strings, which we refer to as *pattern*, and let X be a set of strings, which we refer to as *text*. $|W|$ and $|X|$ denote the cardinalities, i.e., number of elements, of W and X, respectively. The pattern *occurs* in the text if and only if at least one string in W occurs in at least one string in X. The number of occurrences of W in X is given by the sum of the number of occurrences of each of its strings in the strings in X. The z-score z_W for W w.r.t. X and a given background model $\mathcal{M}$ for the data is defined as follows:

$$z_W = \frac{N_W - E}{\sigma}, \tag{15.1}$$

where:

- N_W is the number of occurrences of W in X.

- E is the expected value of the random variable number of occurrences of W in a finite set of strings, each of finite length, generated from $\mathcal{M}$.

- σ is the standard deviation of the same random variable.

We now provide additional details about $\mathcal{M}$, E and the variance σ^2.

15.2.1 A Markov Chain as a Background Model

We use a *Markov chain* (MC, for short) of finite order m to formalize the background model for the data. It is worth recalling (Durbin, Sean, Krogh, and Mitchison 1999; Rabiner 1989) that an MC induces a probability distribution P on Σ^*. Moreover, it can be seen either as a generator of strings according to P or as a tool to evaluate the probability of a string, again according to P. We take this second viewpoint. In our setting, MC can either be known *a priori* or it can be learned from a training set S. In the latter case, S must be a good representative of the statistical properties of the information source one is trying to model, i.e., a *typical set* for the source.

For convenience, from now on, M denotes an MC of order m. For future reference, we point out that an instantiation M of an MC requires either knowledge of the state transition matrix and character stationary distribution or knowledge of a set of strings S from which the mentioned information can be computed. Moreover, in order for M to be used in the computation of the variance and the z-score, additional probability transition matrices and probability vectors are required, which are referred to simply as *auxiliary information*. They are also computed as soon as M is defined.

15.2.2 Expected Value

Using linearity of expectation, the expected value of a multi-set of strings W w.r.t. to M and X is given by the formula:

$$E = \sum_{i=1}^{|W|} E_i, \tag{15.2}$$

where E_i is the expected value of w_i, the i-th string in W. Once again due to the linearity of the expectation, each E_i is (Sinha 2002):

$$E_i = \sum_{h=1}^{|X|} \sum_{j=1}^{n_h - l_i + 1} p_j(c) \cdot p_*(w_i),$$

where c is the first character of w_i, l_i is the length of w_i and n_h is the length of the string x_h in X. Moreover, $p_*(w_i)$ is the probability that w_i occurs in any position of x_h and it is computed via the state transition matrix of M. The term $p_j(c)$ is the probability that c occurs at position j of x_h. Assuming that $p_j(c)$ is independent of j, i.e., assuming that M has a *character stationary distribution p* (of the occurrence of the first symbol of w_i in a string generated by M), E_i can be further simplified as follows:

$$E_i = \sum_{h=1}^{|X|} (n_h - l_i + 1) \cdot p(c) \cdot p_*(w_i).$$

This latter formula is the one used here for the computation of E_i. Additional technical details on the derivation of the above formulas can be found in (Kleffe and Borodovsky 1992; Sinha and Tompa 2000).

For later reference, we point out that a method computing E must take as input a multi-set of strings W, information about the cardinality of the text X and the length of each string in it, and a suitable encoding of M.

15.2.3 Variance

Following Sinha and Tompa (2000), the computation of the variance can be conveniently broken down into the computation of three quantities, as follows:

$$\sigma^2 = B + 2C - E^2, \tag{15.3}$$

E is the expected value of W, as specified above. Intuitively, the term B accounts for the autocorrelation of the strings in W, with respect to their occurrences in strings in X. C is analogous to B, but it accounts for correlation instead of autocorrelation. Computation of all those quantities critically depends on M.

For later reference, as in the case of E, a method computing σ^2 must take as input a multi-set of strings W, information about the text X, and a suitable encoding of M.

15.2.4   The Special Case of DNA

When W and X are meant to be DNA sequences, i.e., strings over the alphabet $\Sigma = \{A, C, G, T\}$ satisfying the well-known base-pairing and complementarity rules, the computation of the z-score must be suitably modified. Indeed, the strings in W *encode* pieces of DNA strings on the same strand of the genomic sequences one is analyzing, those latter being represented by strings in X. However, accounting only for the occurrences of W in X results in an overlook of the matches induced by W on the other DNA strand, yielding an underestimation of the quantities involved in Formula (15.1). In order to carry out a correct computation, one has to include in the pattern multi-set also the reverse complements of each string in W. Let W' be this new multi-set. However, now an *over-counting problem* crops up. As pointed out by Sinha and Tompa, such a problem involves semi-palindromes. A *semi-palindrome* is a string w that can be superimposed with, i.e, it is equal to, its reverse complement, as for instance $AATT$. If a semi-palindrome were in W, it would appear also in W' in the form of its reverse complement and that would lead to count twice for the occurrences of that string in any text, including the ones generated from M to compute the variance. More precisely, such an overcounting would affect term B in Formula (15.1). In conclusion, in the case of DNA, one has to modify the computation of B taking into account semi-palindromes. Additional details can be found in the already mentioned paper by Sinha and Tompa.

15.3 SeqAn Algorithms and Data Types

We now describe the data types and functions that are available in SeqAn for the computation of the statistical indices discussed in the previous section, also for the case of DNA.

15.3.1 The Data Type `MarkovModel`

This data type gives a suitable representation of an MC and provides functions for its instantiation and use in SeqAn; see Table 37. We first describe these functions and illustrate their use with some examples. Then, in Section 15.4, we give some details about the internal representation of the data type.

Listing 32 illustrates how to build and store an MC. A training set composed of N strings is taken as input in lines 1-7. The remaining part of the listing is dedicated to declare, build and store a `MarkovModel` M of order $m = 3$.

Listing 33 shows how to read a previously stored MarkovModel M (lines 20-21), how to input a set of three strings W (lines 23-26), how to compute the probability of the entire set W (line 28) and that of a single string (line 30), with respect to M.

15.3.2 The Functions `expectation`, `variance` and `zscore`

Listing 34 shows an example where the two functions **expectation** and **variance** are used in order to compute the expectation and the variance of the multi-set W w.r.t. M and X. In addition to the inputs of the previous functions, **zscore** needs an indication of which string matching algorithm to use in order to compute the number of occurrences of the pattern in the text, among the many algorithms provided by SeqAn (see Table 19 on page 147). Listing 35 shows how the function may be exploited, with use of two different string matching algorithms.

buildMarkovModel	Given the training set S, this function computes the transition matrix, the character stationary distributions and the auxiliary information for the desired instance M of MC.
setMarkovModel	Given the transition matrix and, possibly, the character stationary distributions of M, this function creates a new instance of MarkovModel corresponding to that information. The auxiliary information is also computed. If the character stationary distributions of M are not provided, it is computed from the transition matrix.
emittedProbability	This function stores in a text file F the transition matrix, the character stationary distributions and the auxiliary information of M.
write	Given a text file F, containing a proper representation of a data type MarkovModel as returned by the function write, read loads it into a corresponding instance M of MC.
read	Given a multi-set of strings W as input, this function computes the cumulative probability P for W w.r.t. M. It is $P = \sum_i^{|W|} P(w_i)$, where $P(w_i)$ is the probability that w_i is emitted by M.

Table 37: **Markov Model Functions.**

```
1   typedef char TAlphabet;
2   typedef String<TAlphabet> TSequence;
3   StringSet<TSequence> S;
4   appendValue(S,str_1);
5   appendValue(S,str_2);
6   ...
7   appendValue(S,str_N);

8   unsigned int m =3;
9   // define a MarkovModel M of order m
10  MarkovModel<TAlphabet> M(m);

11  // build M w.r.t. the training set S
12  buildMarkovModel(M, S);

13  FILE *fd = fopen("Model.txt", "w+");
14  // store M into the file "Model.txt"
15  write(fd, M);
16  fclose(fd);
```

Listing 32: **Instantiation and Storage of an MC.**

```
17   MarkovModel<TAlphabet> M(m);
18   fd = fopen("Model.txt", "r");

19   // load from the file ''Model.txt'' the MarkovModel M
20   read(fd, M);
21   fclose(fd);

22   // input a set of three strings W
23   StringSet<TSequence> W;
24   appendValue(W, string1);
25   appendValue(W, string2);
26   appendValue(W, string3);

27   // output: the probability that set W is emitted by M
28   std::cout << emittedProbability(M, W);

29   // output: the probability that string1 is emitted by M
30   std::cout << emittedProbability(M, string1);
```

Listing 33: **Example Program** for `MarkovModel`. The program reads M and a set of strings W. Then, it computes the probability of W and that of a single string, according to M.

```
1   /* compute expectation of the multi-set of patterns W w.r.t.
2   the text X and the MarkovModel M */

3   TFloat E = expectation(W, X, M);


4   /* compute variance of the  multi-set of patterns W w.r.t.
5   the text X and the MarkovModel M */

6   TFloat V = variance(W, X, M);
```

Listing 34: **The Primitives** expectation and variance.

```
7   /* Compute the z-score for the set of pattern W w.r.t. the
8   text X and the MarkovModel M using the AhoCorasick algorithm
9   to compute the number of occurrence of W in X */
10  TFloat z1 = zscore<AhoCorasick>(W,X,M);

11  /* Compute the z-score using the WuManber algorithm
12  instead of the AhoCorasick one*/
13  TFloat z2 = zscore<WuManber>(W,X,M);
```

Listing 35: **Use of zscore.**

15.3.3   The Special Case of DNA

As mentioned in Section 15.2.4, the statistical indices computation has to be slightly modified for the case of DNA. The affected functions are **variance** and **zscore**. In order to make them work correctly one has simply to specify that the alphabet for the strings is that of DNA, i.e., one uses **typedef DNA TAlphabet** instead of, for example, **typedef char TAlphabet** used in Listing 32.

15.4   Implementation Outline

We now briefly discuss some implementation details concerning the data type **MarkovModel** included in SeqAn. Recall from the previous section that the functions **build** and **set**, although they take as input different things, give the same output. In particular, in both cases, the auxiliary information required for the use of M in the computation of variance and z-score is computed with the function **computeAuxiliaryMatrices**, which is not public.

From now on, we concentrate on the format of **MarkovModel**. Since SeqAn provides utilities that are optimized for the management of string data, matrices and arrays have been conveniently stored and managed as strings. Figure 47 illustrates the format of the file representing a Markov model. Such a format is expected by **read** and it is enforced by **write**. In particular, the terms $a_{i,j}$

and b_j represent the elements of the transition matrix and of the stationary distribution, respectively; $c_{i,j}$, $d_{i,j}$ and $e_{i,j}$ are the terms of the auxiliary matrices needed for the computation of variance and z-score.

$a_{1,1}$	\<space\>	$a_{1,2}$	\<space\>	...	\<space\>	$a_{1,\mid\Sigma\mid^m}$	\<return\>
$a_{2,1}$	\<space\>	$a_{2,2}$	\<space\>	...	\<space\>	$a_{2,\mid\Sigma\mid^m}$	\<return\>
⋮							
$a_{\mid\Sigma\mid^m,1}$	\<space\>	$a_{\mid\Sigma\mid^m,2}$	\<space\>	...	\<space\>	$a_{\mid\Sigma\mid^m,\mid\Sigma\mid^m}$	\<return\>
b_1	\<space\>	b_2	\<space\>	...	\<space\>	$b_{\mid\Sigma\mid^m}$	\<return\>
$c_{1,1}$	\<space\>	$c_{1,2}$	\<space\>	...	\<space\>	$c_{1,\mid\Sigma\mid^m}$	\<return\>
$c_{2,1}$	\<space\>	$c_{2,2}$	\<space\>	...	\<space\>	$c_{2,\mid\Sigma\mid^m}$	\<return\>
⋮							
$c_{\mid\Sigma\mid^m,1}$	\<space\>	$c_{\mid\Sigma\mid^m,2}$	\<space\>	...	\<space\>	$c_{\mid\Sigma\mid^m,\mid\Sigma\mid^m}$	\<return\>
$d_{1,1}$	\<space\>	$d_{1,2}$	\<space\>	...	\<space\>	$d_{1,\mid\Sigma\mid^m}$	\<return\>
$d_{2,1}$	\<space\>	$d_{2,2}$	\<space\>	...	\<space\>	$d_{2,\mid\Sigma\mid^m}$	\<return\>
⋮							
$d_{\mid\Sigma\mid^m,1}$	\<space\>	$d_{\mid\Sigma\mid^m,2}$	\<space\>	...	\<space\>	$d_{\mid\Sigma\mid^m,\mid\Sigma\mid^m}$	\<return\>
$e_{1,1}$	\<space\>	$e_{1,2}$	\<space\>	...	\<space\>	$e_{1,\mid\Sigma\mid^m}$	\<return\>
$e_{2,1}$	\<space\>	$e_{2,2}$	\<space\>	...	\<space\>	$e_{2,\mid\Sigma\mid^m}$	\<return\>
⋮							
$e_{\mid\Sigma\mid^m,1}$	\<space\>	$e_{\mid\Sigma\mid^m,2}$	\<space\>	...	\<space\>	$e_{\mid\Sigma\mid^m,\mid\Sigma\mid^m}$	\<return\>

Figure 47: **Markov Model File Format.**

Authors' Contributions and Acknowledgements

Simona E. Rombo and Filippo Utro contributed equally in the design and implementation of the functions. Raffaele Giancarlo coordinated the research. All three authors contributed equally to the writing of the manuscript. Raffaele Giancarlo and Filippo Utro are partially supported by MIUR-FIRB International Project "Scoperta di Patterns in Strutture Discrete, con Applicazioni alla Bioinformatica."

Chapter 16

A BWT-Based Suffix Array Construction

Tobias Marschall*, Marcel Martin* and Sven Rahmann*

In this chapter, we offer a different perspective on using SeqAn as a research tool and ask, how easily can SeqAn itself be extended? We do not want to plug together existing components, but write a component ourselves. We choose a relatively unknown suffix array construction algorithm called BwtWalk.

Choosing SeqAn as an implementation platform gives us various benefits. First, we can easily compare the new algorithm to the existing suffix array construction algorithms that have already been implemented in SeqAn. Second, by using the same standard interface as those other algorithms, our algorithm can be used in all situations in which an arbitrary suffix array construction algorithm is required. And since the algorithm has been integrated into SeqAn, the implementation is available to the community.

We describe the ideas behind BwtWalk, how we extend SeqAn with its implementation, and how we use SeqAn to test and benchmark it. The reader should be familiar with the suffix array basics described in Chapter 11.

16.1   Introduction to BwtWalk

We call the algorithm BwtWalk because it simulates walking along (partial) Burrows-Wheeler transform (see below) of the in-

---

*Bioinformatics for High-Throughput Technologies, Computer Science 11, TU Dortmund, Germany.

put string s to construct the suffix array. It is well known that s can be reconstructed from its BWT $\hat{s}$ and that $\hat{s}$ is generally easier to compress than s (for suitably structured non-random s).

Most programs that use the BWT (e.g., for compression) first construct the suffix array of s explicitly and then obtain the BWT from it, a prominent example being Julian Seward's `bzip2` compressor[1]. In contrast, the method presented here uses BWTs of successively longer suffixes of s to construct a linked suffix list of s and from it the suffix array. The method is easy to understand and implement, quite efficient on small alphabets and – with some engineering – allows an in-place variant that we describe as well. The worst-case complexity of the standard method, which we call BwtWalkFast, is $\Theta(n \log n)$ on strings of length n, but we observe linear complexity in practice, except on artificial inputs; for the in-place method, which we call BwtWalkInPlace, the worst-case complexity is $\Theta(n^2)$. The extra memory usage (in addition to the space for the input string and the suffix array) is n words for BwtWalkFast, and no extra memory (or n bits if one is picky) for BwtWalkInPlace.

The BwtWalk idea was independently developed first by Baron and Bresler from 2002 onwards and later by the authors (from 2005 onwards) before the publication of Baron and Bresler's paper (Baron and Bresler 2005), which was pointed out to us by Karsten Klein (TU Dortmund). While Baron and Bresler engineered an asymptotically faster implementation using hierarchical lists with $\Theta(n\sqrt{\log n})$ worst-case time but significantly higher memory usage (about $3n$ extra words), we focus on the in-place method BwtWalkInPlace, which is, to our knowledge, unpublished.

---

[1] `http://www.bzip.org/`

16.2 The Main Idea of BwtWalk

In what follows, Σ is an alphabet of constant size, $s = s_0 \ldots s_{n-1} = s[0] \ldots s[n-1]$ is a string of length n over Σ, and $s^p := s_p \ldots s_{n-1}$ denotes the suffix starting at position p. In contrast to Chapter 11, we call the suffix array pos (and not S or SA) and its inverse rank (and not I), because pos[r] is the starting *position* of the lexico-graphically r-th smallest suffix and rank[p] is the lexicographic *rank* of suffix s^p.

The BWT of s is a pair $(\widehat{s}, f)$ consisting of a permutation $\widehat{s}$ of the characters of s and the position f of the last character of s in $\widehat{s}$. If s ends with a unique character, the value of f can be inferred directly from $\widehat{s}$ and is therefore redundant; in this case, only $\widehat{s}$ is called the BWT of s. The BWT is defined by $\widehat{s}[r] := s[\text{pos}[r] - 1]$, where $s[-1]$ is to be understood as $s[n-1]$. In other words, $\widehat{s}[r]$ is the character that appears in s immediately before the starting position of the lexicographically r-th smallest suffix.

We will construct a linked suffix list (a representation of the suffix array) of s in n rounds, starting with round $n-1$, counting back-wards, and finishing with round 0. After round p, we have a suffix list of s^p.

For $0 \le p < n$, we define lexnextpos[p] := pos[rank[p] + 1], the starting position of the suffix that comes next in lexicographic order after s^p. If such a suffix does not exist (because rank[p] = $n - 1$), we set lexnextpos[p] := $\perp$ (a special nil value). Similarly, we define lexprevpos[p] := pos[rank[p] − 1]. With lexnextpos and lexprevpos, we thus simulate a doubly linked list that keeps the suffix starting positions in lexicographic order in each round. For each $c \in \Sigma$, we further define lexfirstpos[c] and lexlastpos[c] as the starting positions of the lexicographically first and last suffixes that begin with c. They facilitate inserting suffixes $s^p = cu$, where $c \in \Sigma$ does not occur in $u \in \Sigma^*$. Initially (before round $n-1$), we set lexfirstpos[c] = lexlastpos[c] = $\perp$ for all $c \in \Sigma$.

At the beginning of round p, the arrays lexprevpos[$p+1, \ldots, n-1$], lexnextpos[$p+1, \ldots, n-1$], lexfirstpos and lexlastpos represent the suffix array of s^{p+1}. There are two different cases for the update in round p, based on s_p.

(1) The easy case occurs if s_p does not occur in s^{p+1}, that is, if and only if $\texttt{lexfirstpos}[s_p] = \texttt{lexlastpos}[s_p] = \bot$. We then need to find the largest character $c^- < s_p$ for which $p^- :=$ $\texttt{lexlastpos}[c^-] \neq \bot$ and the smallest character $c^+ > s_p$ for which $p^+ := \texttt{lexfirstpos}[c^+] \neq \bot$. Since presently p^- precedes p^+ in lexicographic order, we only need to insert p between p^- and p^+ by updating both $\texttt{lexnextpos}$ and $\texttt{lexprevpos}$ accordingly. We also initialize $\texttt{lexfirstpos}[s_p]$ and $\texttt{lexlastpos}[s_p]$ to p.

(2) If s_p already occurs somewhere in s^{p+1}, we need to determine the correct positions p^- and p^+, between which p must be inserted to maintain lexicographic order. They are found with the following lemma: If there exists a position $q > p$ with $s_p = s_q$, and s^{q+1} is the direct lexicographic predecessor (successor) of s^{p+1}, then s^q is the direct lexicographic predecessor (successor) of s^p. This is seen by a contradiction argument and leads to the following algorithm to find the direct lexicographic predecessor position if it exists.

Starting at position $p + 1$ (of the most recently inserted suffix), follow the $\texttt{lexprevpos}$ links to the left in the simulated linked list. Thus, let $c := s_p$ be the character to be searched for, start at $i \leftarrow$ $\texttt{lexprevpos}[p + 1]$, and check whether $s_{i-1} = c$. If not, continue following the $\texttt{lexprevpos}$ links ($i \leftarrow \texttt{lexprevpos}[i]$). If eventually $s_{i-1} = c$, insert p into the list after $p^- := i - 1$. In other words, writing $p^+ := \texttt{lexnextpos}[p^-]$, we update $\texttt{lexnextpos}[p^-] \leftarrow p$ and $\texttt{lexnextpos}[p] \leftarrow p^+$, and we update $\texttt{lexprevpos}$ accordingly. Additionally, if previously $\texttt{lexlastpos}[c] = p^-$, we must update $\texttt{lexlastpos}[c] \leftarrow p$. On the other hand, if we eventually fall off the list ($i = \bot$), then s^p must be the lexicographically first suffix starting with c, and we update the arrays accordingly.

Indeed, the above paragraph describes the main loop of the algorithm. Comparing s_{i-1} to c for different i can be visualized as walking leftward along the BWT while searching for the first occurrence of c; hence the name BWTWALK.

To complete the description of the BWTWALKFAST algorithm, it remains to say that we can similarly walk rightward in the BWT (following the $\texttt{lexnextpos}$ links), and we can do so in both directions in parallel, stopping at the first c in either direction. Interestingly, as shown by Baron and Bresler (2005), this small measure improves the worst-case running time from $\Theta(n^2)$, when walking only leftward or only rightward, to $\Theta(n \log n)$.

To create the suffix array `pos` from the suffix list, we start with $r \leftarrow 0$ at position $p \leftarrow$ `lexfirstpos`$[a]$, where a is the smallest character, and repeatedly set `pos`$[r] \leftarrow p$ and increment r, while following the links ($p \leftarrow$ `lexnextpos`$[p]$). Because `lexprevpos` is no longer required, it can occupy the same memory as `pos` (if `pos` allows fast random access; see Section 16.5); so we need n extra words for `lexnextpos`. The following section explains how to avoid using this extra space.

16.3   Saving Space

We can save space by storing both `lexprevpos` and `lexnextpos` in the *same* array using an old folklore trick by xor-ing their values, i.e., we only use a single array

$$\texttt{lexxorpos}[p] := \texttt{lexprevpos}[p] \ \texttt{xor} \ \texttt{lexnextpos}[p].$$

If either `lexprevpos`$[p]$ or `lexnextpos`$[p]$ is known, the other value can be computed because the xor-operation is its own inverse.

It is still possible to "walk" unidirectionally through the array if the first or last value is known. However, it is not informative to randomly access `lexxorpos`$[p]$; from this value alone, neither previous nor next position can be determined.

Therefore, to insert p between p^- and p^+, as described for BWT-WALKFAST, it is not sufficient to determine either p^- or p^+ by walking along the BWT, because it is now impossible to find the other value by following a `lexnextpos` or `lexprevpos` link. Instead, both p^- and p^+ need to be determined by walking until the sought character is found in both directions. Depending on the structure of the input string, this sometimes only approximately doubles the running time, but can have a much stronger effect and increases the worst-case complexity to $\Theta(n^2)$ (see Section 16.7).

After constructing `lexxorpos`, another complication arises: We cannot simply overwrite it with `pos` in-place. This is not a problem if `pos` is an external file because it can be written sequentially to disk, while `lexxorpos` remains in memory. However, if `pos` is desired or required to be in memory, we may proceed as follows (suggested by David Weese): Following the (implicit) `lexnextpos` links, we overwrite each visited element with sequential numbers. This overwrites `lexxorpos` with the inverse suffix array `rank` in place. It remains to overwrite `rank` with its inverse. Fortunately, Knuth's *The Art of Computer Programming,* Vol. 1, Section 1.3.3, contains an "Algorithm I" (Knuth 1997) that performs this task (almost) in place, using the cycle structure of the permutation. However, an additional bit per element (e.g., the sign bit) is required to mark already visited elements. If $n < 2^{31}$ (on 32-bit architectures, resp. $n < 2^{63}$ on 64-bit architectures), this does not consume additional memory in practice. Otherwise, we explicitly allocate n bits temporarily.

16.4 SeqAn Implementation of BWTWALKFAST

16.4.1 Getting Started

Among others, SeqAn contains a simple suffix array construction algorithm that uses quicksort (`seqan/index/index_sa_qsort.h`), which serves as a guide.

In SeqAn, functionality is offered through *global template functions* whenever possible (see Section 4.4). Different algorithms that solve the same problem are differentiated by *tags* (see Section 4.6.2). At first, we implement the BWTWALKFAST algorithm. We therefore define the tag `BwtWalkFast`:

```
struct _BwtWalkFast {};
typedef Tag<_BwtWalkFast> const BwtWalkFast;
```

We follow the coding convention to prepend an underscore to names meant to be used only locally. This coding practice simulates private declarations in SeqAn (see Section 4.4.2). Having the tag in place, we overload the global template function `createSuffixArray`:

```
template < typename TSA, typename TText >
void createSuffixArray(TSA &SA, TText &s,
                              BwtWalkFast const &)
```

The template parameters `TText` and `TSA` specify the type of the input text and the type of the suffix array to be computed, respectively. We could, for example, use `String<Dna>` for `TText` and `String<unsigned>` for `TSA`. To make our implementation as generic as possible, we derive other needed types from these template parameters. Table 38 summarizes all used types, whose precise definitions are given at the appropriate places throughout the next sections.

TAllowsFastRandomAccess	Defined either as `True` or `False`, depending on the `AllowsFastRandomAccess` metafunction.
TAlphabetArray	Type of `lexfirstpos` and `lexlastpos` arrays.
TArray	Type of `lexprevpos` and `lexnextpos` arrays; always in internal memory.
TChar	Type of a text character.
TSA	Type of the suffix array; usually `String<unsigned>` or `String<unsigned long long>`.
TSaIter	Iterator over the suffix array.
TText	Type of the input text.
TTextIter	Iterator over the input text.
TValue	Type of a suffix array entry.

Table 38: **Overview of all Template Type Parameters and Derived Typedefs.**

The algorithm can naturally be divided into several functions. This improves clarity but does not compromise performance as the functions are inlined by the compiler when appropriate.

16.4.2   List Operations

As explained in Section 16.2, we use two arrays `lexnextpos` and `lexprevpos` to represent the linked suffix list. Within the lists, a special constant named `NIL` represents invalid list indices $\perp$. With the help of the `SupremumValue` metafunction (see Section 4.5), `NIL` is defined to be the largest representable value of the index data type `TValue`.

```
const TValue NIL = SupremumValue<TValue>::VALUE;
```

We can now understand the first helper function, which inserts a value `p` into a list:

```
template < typename TArray, typename TValue >
void _insertBetween(TArray &lexprevpos,
    TArray &lexnextpos, TValue p,
    TValue predecessor, TValue successor)
{
   value(lexprevpos, p) = predecessor;
   value(lexnextpos, p) = successor;
   const TValue NIL = SupremumValue<TValue>::VALUE;
   if (predecessor != NIL)
      value(lexnextpos, predecessor) = p;
   if (successor != NIL)
      value(lexprevpos, successor) = p;
}
```

Note how we follow the SeqAn policy of using the `value` function instead of the index operator (see Section 5.1.4).

For insertion, both **predecessor** and **successor** need to be known. For convenience, we provide the two functions `_insertAfter` and `_insertBefore` that simply look up predecessor or successor in `lexnextpos` or `lexprevpos`, respectively, before calling `_insertBetween`. If **predecessor** and **successor** are already known, we can still call _insertBetween directly and save one table look-up.

16.4.3   Main Algorithm

Using the helper functions defined in the previous section, the main part of our algorithm can be written down concisely. We declare the function _createSuffixList, which contains the algorithm's main building block, namely the construction of the linked list representation of the suffix array. It returns the position of the smallest suffix, or, in other words, the head of the list. Let us take a look at the function declaration and useful typedefs:

```
template < typename TArray, typename TText >
typename Value<TArray>::Type _createSuffixList(
    TArray &lexprevpos, TArray &lexnextpos, TText &s)
{
    typedef typename Iterator<TText, Standard>::Type
                                                TTextIter;
    typedef typename Value<TArray>::Type TValue;
    typedef typename Value<TText>::Type TChar;
    const unsigned int ALPHABETSIZE =
                                ValueSize<TChar>::VALUE;
    const TValue NIL = SupremumValue<TValue>::VALUE;
    // ... continued below
```

More metafunctions are used here: **Value** determines the type of elements in a container, whereas **Iterator** returns an appropriate iterator to traverse the container; by specifying **Standard** as second template argument, we communicate that we do not need a rooted iterator (see Section 6.5). The third metafunction, **ValueSize**, allows us to conveniently determine the alphabet size.

Our next concerns are allocation and initialization of the auxiliary arrays lexfirstpos and lexlastpos. Table 6 on page 85 tells us that an array string is the right data type to use:

```
typedef String<TValue,Array<ALPHABETSIZE> >
                                  TAlphabetArray;
TAlphabetArray lexfirstpos, lexlastpos;
fill(lexfirstpos, ALPHABETSIZE, NIL, Exact());
fill(lexlastpos, ALPHABETSIZE, NIL, Exact());
```

Here, the `fill` function is used to resize and fill the newly created array strings in one step. Since we know that the arrays will not grow, we use the tag **Exact** to reserve only the required amount of memory (and not more).

We now have all needed types and arrays in place and proceed to the algorithm's main part. In each iteration, we insert one suffix into the list of suffixes, starting with the shortest suffix. To iterate over those suffixes, we use the iterators provided by SeqAn (see Section 7.5):

```
TTextIter it = end(s);
TValue p = length(s);
while (it != begin(s)) {
    --it;
    --p;
    TChar c = *it;
    unsigned int cOrd = ordValue(c);

    // main step: insert suffix that starts at position p
    // ...
}
```

The used `ordValue` function maps each character of an alphabet to a positive integer (see Section 6.4 for details), allowing us to use `cOrd` as an index to the `lexfirstpos` and `lexlastpos` arrays. To complete the `_createSuffixList` function, only the insertion operation needs to be implemented. The corresponding code fragment is shown in Table 36. Recall from Section 16.2 that we need to differentiate between two cases. If we see the current character `c` for the first time, `lexfirstpos` and `lexlastpos` provide the correct insertion position and need to be updated accordingly (code omitted). Most of the time, the character has already occurred at

least once. In this case, we walk along the preliminary BWT until
we find the character c. If we hit the beginning (end), we know
the current suffix to be the first (last) suffix starting with c and
can insert it accordingly (code omitted).

```
// first time we see character c?
if (value(lexfirstpos, cOrd) == NIL) {
  // ... omitted
} else
{
  TValue pLeft  = value(lexprevpos, p+1); // walking left
  TValue pRight = value(lexnextpos, p+1); // walking right
  while (true) {
    // end of list found while walking left?
    if (pLeft == NIL) {
      // ... omitted
     break;
    } else
    {
      // character found walking to the left?
      if (value(s, pLeft-1) == c) {
        _insertAfter(lexprevpos, lexnextpos, p, pLeft-1);
        if (value(lexlastpos, cOrd) == pLeft-1)
          value(lexlastpos, cOrd) - p;
        break;
      }
    }
    pLeft = value(lexprevpos, pLeft);
    // ... (analogous code for walking right omitted)
  }
}
```

Listing 36: **The Main Loop of** `_createSuffixList`.

This completes the function **`_createSuffixList`**. Given the suffix
list, we still need to build the suffix array. We address this issue in
the next section. Furthermore, we see how the arrays **lexprevpos**
and **lexnextpos**, which we assumed to be given, can smartly be
provided.

16.5   Containers with and without Fast Random Access

When `createSuffixArray` is called, the caller provides a reference to the suffix array to be filled:

```
template < typename TSA, typename TText >
void createSuffixArray(TSA &SA, TText &s,
                              BwtWalkFast const &)
```

Temporarily storing the `lexprevpos` array in the memory provided by `&SA` (see Section 16.2) is only sensible if `TSA` refers to a type that allows fast random access. We cannot take this for granted as `TSA` could be an external string type (see Section 7.3.5). Fortunately, this issue can be resolved by using the `AllowsFastRandomAccess` metafunction.

```
template < typename TSA, typename TText >
void createSuffixArray(TSA &SA, TText &s,
                              BwtWalkFast const &)
{
  typedef typename AllowsFastRandomAccess<TSA>::Type
                              TAllowsFastRandomAccess;
  _createSuffixArray(
      SA, s, BwtWalkFast(), TAllowsFastRandomAccess());
}
```

We use `createSuffixArray` as a wrapper and move the actual implementation into the function `_createSuffixArray`, of which we provide two overloaded variants. The result returned by the metafunction `AllowsFastRandomAccess` is either the type `True` or the type `False`. Hence, the signatures of the two functions are the following:

```
template < typename TSA, typename TText >
void _createSuffixArray(
    TSA &SA, TText &s, BwtWalkFast const &, True const &)

template < typename TSA, typename TText >
void _createSuffixArray(
    TSA &SA, TText &s, BwtWalkFast const &, False const &)
```

In the first variant, we use the provided SA memory for `lexprevpos` and allocate additional memory only for `lexnextpos`, while in the second one, we allocate memory for both `lexprevpos` and `lexnextpos`. The latter variant only accesses SA once and only consecutively when creating the suffix array from the suffix list. Both variants call `_createSuffixList` and traverse the generated list in order to build the suffix array:

```
// variant-specific array allocation code omitted
TValue p = _createSuffixList(lexprevpos, lexnextpos, s);
TSaIter saIt = begin(SA);
while (p != NIL) {
  *saIt = p;
  p = value(lexnextpos, p);
  ++saIt;
}
```

16.6   In-Place Version

As explained in Section 16.3, BWTWALKFAST's space consumption can be reduced at the cost of additional running time by combining `lexprevpos` and `lexnextpos` into a single array `lexxorpos`. Now we are in a frequently encountered situation: There are two versions of an algorithm with different space-time trade-offs, and we want to let the user decide which version to use. SeqAn's answer to this situation is *template subclassing* (see Section 4.3):

```
struct _BwtWalkFast {};
typedef Tag<_BwtWalkFast> const BwtWalkFast;

struct _BwtWalkInPlace {};
typedef Tag<_BwtWalkInPlace> const BwtWalkInPlace;

template <typename TSpec = BwtWalkFast >
struct BwtWalk {};
```

Here, we declare a tag BwtWalk for the whole *class* of BWT-WALK algorithms. The tag has a template parameter. By filling in `BwtWalkFast` or `BwtWalkInPlace`, one can choose the desired version, called algorithm *subclass*. We modify the main wrapper function `createSuffixArray` to pass on the subclass parameter:

```
template < typename TSA, typename TText, typename TSpec >
inline void createSuffixArray(
   TSA &SA, TText &s, BwtWalk<TSpec> const &)
{
   typedef typename AllowsFastRandomAccess<TSA>::Type
      TAllowsFastRandomAccess;
   _createSuffixArray(
      SA, s, BwtWalk<TSpec>(), TAllowsFastRandomAccess());
}
```

In total, we provide four overloaded versions of `_createSuffixArray`: The user may choose between `BwtWalkFast` and `BwtWalkInPlace`, and at the same time the provided container may allow random access or not. Although we provide four versions of the same function, almost no code is duplicated. These functions are short since most of the work is done in `_createSuffixList` or `_createXoredSuffixList`, respectively. The latter function contains the main part of the algorithm using the space efficient `lexxorpos` array. Recall that, in this case, we have to walk along the BWT into both directions (code omitted).

16.7 Experiments

Implementing our algorithms within SeqAn allows us to easily compare them to the existing ones. In this section, we show the results of this comparison.

16.7.1 A Demo Program

It is time to see how to read a text file, run a suffix array construction algorithm, and measure its running time in SeqAn. Listing 37 shows a code example.

```
1   #define SEQAN_PROFILE
2   #include <fstream>
3   #include <seqan/index.h>
4   #include "index_sa_bwtwalk.h"

5   using namespace seqan;
6   using namespace std;

7   int main() {
8       fstream inputFile("example.txt");
9       CharString text;
10      read(inputFile, text, Raw());
11      String<unsigned> sa;
12      resize(sa, length(text));
13      SEQAN_PROTIMESTART(time);
14      createSuffixArray(sa, text,
15          BwtWalk<BwtWalkFast>());
16      cout << SEQAN_PROTIMEDIFF(time) << endl;
17      return 0;
18  }
```

Listing 37: **Example for** Bwt Walk Fast. This small example program that runs the Bwt Walk Fast algorithm on the file **example.txt** and prints the time needed to compute the suffix array.

Note the definition of the `SEQAN_PROFILE` preprocessor macro in line 1. The macros for time measurements are only available when this macro is defined. When starting time measurements, `SEQAN_PROTIMESTART` is used. It stores the current time in a variable whose name is given as a macro parameter. To read out the elapsed time, `SEQAN_PROTIMEDIFF` is used as if it was a function that returns a floating point value representing the number of seconds since the start of the timer. The measured time is wall-clock time.

16.7.2   Experimental Setup

For our experiments, we use a tool that extends the above example. It measures the running time (CPU time, in this case) of several suffix array construction algorithms on various data files. Our corpus is assembled from parts of (1) the Manzini-Ferragina corpus (Manzini and Ferragina 2004)[2], (2) the Gauntlet corpus[3], and (3) the corpus used by Schürmann and Stoye (2007)[4]. Table 39 contains descriptions of all data files.

Since we are also interested in seeing how the algorithms cope with large strings such as the human genome, we include two more files. The first one, `hg18.dna` is created by concatenating the sequences of the NCBI's human genome[5] build 36.1, separated by the $ character, and then converting all characters to upper case. `hg18.seq` was created from that file by replacing the human-readable characters A, C, G, T, N, and $ with the byte values 0, 1, 2, 3, 4, and 255, respectively. Both files therefore have the same length and represent the same information, only the alphabets chosen to encode the sequence differ. The suffix arrays of both files also differ, since the lexicographical order is different: In ASCII encoding, $N < T$, but in byte encoding, $T \equiv 3 < 4 \equiv N$.

---

[2] http://web.unipmn.it/~manzini/lightweight/corpus/

[3] http://www.michael-maniscalco.com/msufsort.htm

[4] http://bibiserv.techfak.uni-bielefeld.de/download/tools/bpr.html

[5] http://hgdownload.cse.ucsc.edu/goldenPath/hg18/bigZips/chromFa.zip

File	(MB)	|Σ|	Description
4Chlamydophila	4.9	6	four *Chlamydophila* genomes
6Streptococci	11.6	5	six *Streptococcus* genomes
A_thaliana_Chr4	12.1	7	*Arabidopsis thaliana* chromosome 4
C_elegans_Chr1	14.2	5	*Caenorhabditis elegans* chromosome 1
E_coli	4.6	4	*Escherichia coli* genome
Fibonacci	20.0	2	Fibonacci string
abac	0.2	3	Gauntlet corpus
abba	10.5	4	Gauntlet corpus
bible	4.0	64	King James bible of the Canterbury Corpus
book1x20	15.4	81	Gauntlet corpus
chr22	34.6	5	*Homo sapiens* chromosome 22
ecoli	14.8	5	The file *E.coli* of the Canterbury corpus
etext	105.3	146	Project Gutenberg texts
fss10	12.1	2	Gauntlet corpus
fss9	2.9	2	Gauntlet corpus
gcc	86.6	150	GCC 3.0 source files (tar archive)
hg18.dna	3107.7	6	See text
hg18.seq	3107.7	6	See text
houston	3.8	16	Gauntlet corpus
howto	39.4	197	Concatenation of Linux Howto files
jdk	69.7	113	.html and .java files from the JDK 1.3
linux	116.3	256	Linux kernel 2.4.5 source files (tar archive)
paper5x80	1.0	92	Gauntlet corpus
period_20	20.0	17	Repeated Bernoulli string
period_1000	20.0	26	Repeated Bernoulli string
period_500000	20.0	26	Repeated Bernoulli string
random	20.0	26	Bernoulli string
reuters	114.7	93	Reuters news in XML format
rfc	116.4	120	Concatenation of RFC text files
sprot	109.6	66	Swissprot database (rel.34)
test1	2.1	256	Gauntlet corpus
test2	2.1	256	Gauntlet corpus
test3	2.1	256	Gauntlet corpus
w3c	104.2	256	HTML files of W3C homepage
world	2.5	94	The CIA world fact book (Canterbury corpus)

Table 39: **Description of Benchmark Data Sets.**

The algorithms that we compare are BwtWalkFast (abbreviated "BwF"), BwtWalkInPlace "BwIP," Manber-Myers "MM," (Manber and Myers 1993) Deep-Shallow "DS," (Manzini and Ferragina 2004), Skew3, (Kärkkäinen and Sanders 2003) and quicksort ("qsort"). There are many more suffix array construction algorithms, some of which are faster than the ones chosen here, but those have—at the time of writing—not yet been incorporated into SeqAn. Also, the primary purpose of this comparison is simply to give an estimate of the performance of the newly implemented algorithms relative to those that already existed in SeqAn.

The experiments were run on a machine that has four AMD Dual-Core Opteron 2.4 GHz processors (i.e., 8 cores) and 128 GB RAM. The algorithms were run one at a time to avoid memory bandwidth issues.

16.7.3 Suffix Array Construction Benchmarks

BwtWalk is an inherently serial algorithm with abysmal cache performance. Each access to `lexnextpos` (and the other arrays) is likely to induce a cache miss in long input texts. Nevertheless, running times in practice are competitive for several strings, especially on small alphabets. All running times are shown in Table 40. BwtWalkFast's specialty is dealing with highly repetitive strings such as abba, abac, Fibonacci, houston, test{1,2,3}, and especially the period_...files. On those strings, BwtWalkFast is always faster than the other algorithms, often by a large margin.

BwtWalkFast is weak for medium to large alphabets. Texts involving the full English alphabet plus perhaps numbers, punctuation, etc. as in the etext, gcc, howto, linux, and w3c files, provoke long running times.

At first, the houston file seems to be an exception. As shown in the table, its alphabet size is medium (16 characters), but Bwt-WalkFast is still the fastest algorithm. On closer inspection, we find that 98% of that file is made up of only three unique characters. In other words, for BwtWalkFast, the distribution of character occurrences is a better indicator of performance than only the alphabet size.

File	DS	BwF	BwIP	Skew3	MM	qsort
4Chlamydophila	2.1	**1.7**	5.2	7.2	29.7	29.4
6Streptococci	**4.0**	4.9	13.2	18.6	90.4	20.6
A_thaliana_Chr4	**3.4**	5.8	11.3	20.1	96.8	19.4
C_elegans_Chr1	**4.0**	6.5	14.2	21.6	107.0	1930.4
E_coli	**1.1**	2.0	3.9	7.0	30.9	4.9
Fibonacci	360.1	**2.6**	8.4	20.6	93.1	—
abac	27.7	**0.0**	17.7	0.0	0.1	903.1
abba	39.7	**1.4**	8.4	12.0	56.0	—
bible	**0.9**	6.2	56.7	6.0	25.7	3.6
book1x20	122.7	**3.4**	1436.2	22.3	79.8	—
chr22	**11.2**	23.4	6595.2	62.7	374.8	3275.6
ecoli	8.5	**4.8**	12.7	24.7	117.1	2767.5
etext	**55.0**	1284.8	7873.5	280.5	1694.0	716.6
fss10	109.9	**1.5**	5.2	12.8	52.7	—
fss9	8.1	**0.3**	1.1	2.6	10.3	10144.0
gcc	**58.4**	540.3	4458.2	179.1	986.7	20524.0
hg18.dna	N/A	**4321.3**	21182.0	16219.5	—	—
hg18.seq	N/A	**4354.5**	7953.1	16131.7	—	—
houston	124.9	**0.5**	201.9	1.3	7.8	8211.0
howto	**13.3**	322.8	1970.7	80.6	421.2	84.3
jdk	**51.8**	146.1	1516.8	134.3	771.6	639.4
linux	**42.6**	634.1	4233.7	262.2	1474.5	509.5
paper5x80	0.8	**0.1**	42.9	0.8	2.7	3237.1
period_20	24785.6	**1.5**	—	19.5	66.5	—
period_1000	537.3	**2.7**	—	28.7	138.9	—
period_500000	362.4	**3.7**	1353.6	33.0	111.2	—
random	**6.0**	43.8	128.6	26.8	200.1	19.7
reuters	**98.8**	437.4	—	262.2	1625.9	642.9
rfc	**48.7**	856.9	6130.2	270.0	1664.3	273.5
sprot	**52.8**	393.3	22227.8	258.3	1580.2	305.2
test1	17.2	**0.2**	—	2.6	13.3	—
test2	17.2	**0.2**	—	2.6	13.3	—
test3	14.8	**0.3**	785.0	2.4	16.4	23001.5
w3c	**83.3**	1019.7	9443.5	217.2	1173.2	14536.9
world	**0.4**	6.0	46.4	3.3	14.4	2.1

Table 40: **Suffix Array Construction Running Times.** A comparison of the running times of some of the suffix array construction algorithms implemented in SeqAn, including BwtWalkFast and BwtWalkInPlace. Dashes (—) denote computations that did not finish within 10 hours; for **hg18**, the limit was increased to 20 hours. Since no 64-bit version of Deep-Shallow is available, it could not be run on the hg18 files ("N/A" in the table). $|\Sigma|$ is the alphabet size. The times include neither reading the input files nor writing the finished suffix array to disk.

On all of the tested data sets, either Deep-Shallow or BwtWalk-Fast is the fastest algorithm. This reflects that Deep-Shallow and BwtWalkFast have opposite strengths. Deep-Shallow is good on large alphabets and bad on repetitive strings, while Bwt-WalkFast is bad on large alphabets and good on repetitive strings.

BwtWalkInPlace is naturally always slower than BwtWalk-Fast. For some sufficiently benign strings, it takes only approximately two or three times as long as BwtWalkFast (ecoli, 4Chlamydophila.dna, C_elegans_Chr1.dna), but in many cases its quadratic worst-case behavior is apparent and it is many times slower than BwtWalkFast.

Fortunately, the human genome sequence hg18.seq is of the benign sort. For creating a suffix array of such huge strings, memory usage becomes a concern and since BwtWalkInPlace is very efficient in that regard, it is still a viable alternative when it is acceptable to trade construction time for memory usage. Interestingly, that trade-off becomes less attractive when simply a different encoding of the alphabet is chosen: Switching from an alphabet that is ordered (A, C, G, T, N) to one that is ordered (A, C, G, N, T) slows down BwtWalkInPlace by a factor of approximately 2.6. Still, both BwtWalk algorithms and Skew3 were the only ones to finish the construction of the hg18 suffix arrays within an acceptable time. Deep-Shallow could not be used for that file since only a 32-bit version is available.

Memory Usage

Since memory consumption is only interesting for large files, we provide it for the construction of the suffix array for hg18.dna. BwtWalkFast needs 49 GB and BwtWalkInPlace needs 26 GB, while Skew3 requires 64 GB. Manber-Myers and the simple quicksort algorithm did not finish their construction of the suffix array, but after some time their memory usage was stable at 78 GB and 26 GB, respectively.

Algorithm	SeqAn Function	Add'l Memory
BwtWalkFast	_createSuffixArrayBwtWalkMin with FastRandomAccess	n words
BwtWalkFast	_createSuffixArrayBwtWalkMin without FastRandomAccess	$2n$ words
BwtWalkInPlace	_createSuffixArrayBwtWalkBoth with FastRandomAccess	none [or n bits]
BwtWalkInPlace	_createSuffixArrayBwtWalkBoth without FastRandomAccess	n words

Table 41: **Overview of Implemented BwtWalk Variants.** See also Sections 16.3 and 16.6.

16.8   Conclusion

We have implemented four versions of the BwtWalk algorithm. All of them need memory for the input text (n characters) and memory for the suffix array (n words), where a word is usually 4 bytes on 32-bit systems (limiting the length of the text that can be indexed) and 8 bytes on 64-bit systems. All versions also need a constant number of extra words for the lexfirstpos and lexlastpos tables and further bookkeeping.

SeqAn allows the output pos array to be on external memory, so the above n words are not required to be available in main memory. However, since the BwtWalk algorithms need fast random access to lexprevpos and to lexnextpos (or to lexxorpos), we can only temporarily use the pos-space for one of the lists if pos allows fast random access. Otherwise, we have to reserve n extra words in main memory. Algorithmically, we have the two variants BwtWalkFast (faster, using n extra words) and BwtWalkInPlace (often only twice as slow, but can be much slower, but uses no extra words of memory). Table 41 gives an overview.

The SeqAn library is fast and generic. This is achieved through a heavy use of templates. A programmer who wants to extend SeqAn first needs to learn its paradigms and conventions. Especially the consequent use of coding conventions is rather important in SeqAn, as some language concepts are replaced by conventions;

for example, private fields are simulated through a prepended underscore and inheritance is replaced by template subclassing. In fact, the first part of this book gives an introduction to these topics and discusses the reasons for these design choices. Once familiar with SeqAn's concepts, it was relatively easy to extend it with a new algorithm. We highly benefited from the fact that a simple suffix array construction algorithm was already available and served as a template.

We hope that this example encourages more researchers and developers to integrate their algorithms into the SeqAn framework, so comprehensive benchmarks on large corpora can be run in the future. Indeed, SeqAn provides an excellent platform to compare different algorithms empirically. The BWTWALKINPLACE example on the human genome shows that large-scale suffix array construction might still offer a few surprises.

We gratefully acknowledge the help of the SeqAn team; especially David Weese and Andreas Döring provided additional ideas and advice. We thank Karsten Klein (TU Dortmund) for pointing us to the Baron and Bresler (2005) paper.

Bibliography

Abouelhoda, M. I., S. Kurtz, and E. Ohlebusch (2002). The enhanced suffix array and its applications to genome analysis. In *WABI '02: Proceedings of the Second International Workshop on Algorithms in Bioinformatics*, pp. 449–463. Springer-Verlag.

Abouelhoda, M. I., S. Kurtz, and E. Ohlebusch (2004). Replacing suffix trees with enhanced suffix arrays. *J. of Discrete Algorithms 2*(1), 53–86.

Abouelhoda, M. I. and E. Ohlebusch (2003). Multiple genome alignment: Chaining algorithms revisited. In *Proceedings of the 14th Annual Symposium on Combinatorial Pattern Matching*, Volume 2676, pp. 1–16. Springer.

Aho, A. V. and M. J. Corasick (1975). Efficient string matching: an aid to bibliographic search. *Communications of the ACM 18*(6), 333–340.

Allauzen, C., M. Crochemore, and M. Raffinot (1999). Factor oracle: A new structure for pattern matching. In *SOFSEM '99: Proceedings of the 26th Conference on Current Trends in Theory and Practice of Informatics on Theory and Practice of Informatics*, pp. 295–310. Springer-Verlag.

Allauzen, C., M. Crochemore, and M. Raffinot (2001). Efficient experimental string matching by weak factor recognition. In *Proceedings of the 12th Annual Symposium on Combinatorial Pattern Matching*, Volume 2089 of *Lecture Notes in Computer Science*, pp. 51–72. Springer.

Altschul, S. F., W. Gish, W. Miller, E. W. Myers, and D. J. Lipman (1990). Basic local alignment search tool. *Journal of Molecular Biology 215*(3), 403–410.

Apostolico, A., M. E. Bock, and S. Lonardi (2003). Monotony of surprise and large-scale quest for unusual words. *Journal of Computational Biology 10*(2/3), 283–311.

Apostolico, A. and C. Pizzi (2007). Motif discovery by monotone scores. *Discrete Applied Mathematics 155*(6-7), 695–706.

Austern, M. H. (1998). *Generic Programming and the STL*. Addison Wesley.

Bailey, T. L. and C. Elkan (1994, August). Fitting a mixture model by expectation maximization to dicover motifs in biopolymers. In *Proceedings of the Second International Conference on Intelligent Systems for Molecular Biology*, pp. 28–36. AAAI Press.

Bailey, T. L. and C. Elkan (1995). The value of prior knowledge in discovering motifs with meme. In *Proceedings of the International Conference on Intelligent Systems for Molecular Biology*, Volume 3, pp. 21–29. AAAI Press.

Baron, D. and Y. Bresler (2005). Antisequential suffix sorting for BWT-based data compression. *IEEE Transactions on Computers 54*(4), 385–397.

Bellman, R. (1957). *Dynamic Programming*. Princeton University Press.

Benson, D. A., I. Karsch-Mizrachi, D. J. Lipman, J. Ostell, and D. L. Wheeler (2008). Genbank. *Nucleic Acids Research 36*, D25–30.

Bentley, D. R. (2006). Whole-genome re-sequencing. *Current Opinion in Genetics and Development 16*(6), 545–552.

Boyer, R. S. and J. S. Moore (1977, October). A fast string searching algorithm. *Communications of the ACM 20*(10), 762–772.

Brudno, M., C. Do, G. M. Cooper, M. F. Kim, E. Davydov, N. C. S. Program, E. D. Green, A. Sidow, and S. Batzoglou (2003). Lagan and multi-lagan: Efficient tools for large-scale multiple alignment of genomic DNA. *Genome Research 13*(4), 721–731.

Buhler, J. and M. Tompa (2001). Finding motifs using random projections. In *RECOMB 2001: Proceedings of the fifth annual international conference on Computational biology*, pp. 69–76. ACM.

Burrows, M. and D. J. Wheeler (1994). A block-sorting lossless data compression algorithm. Technical Report SRC-RR-124, Digital Systems Research Center.

Chen, J. Y. and S. Lonardi (2009). *Biological Data Mining*. Chapman and Hall.

Cormen, T. H., C. E. Leiserson, R. L. Rivest, and C. Stein (2001). *Introduction to Algorithms, second edition*. MIT Press.

Crochemore, M., A. Czumaj, L. Gasieniec, S. Jarominek, W. P. T. Lecroq, and W. Rytter (1994, November). Speeding up two string-matching algorithms. *Algorithmica 12*(4–5), 247–267.

Czarnecki, K. and U. W. Eisenecker (2000). *Generative Programming. Methods, Tools, and Applications*. Addison Wesley.

Darling, A., B. Mau, F. Blattner, and N. Perna (2004). Mauve: Multiple Alignment of Conserved Genomic Sequence with Rearrangements. *Genome Research 14*, 1394–1403.

Davila, J., S. Balla, and S. Rajasekaran (2006). Space and time efficient algorithms for planted motif search. In *IWBRA 2006: Second International Workshop on Bioinformatics Research and Applications*, Volume 3992 of *Lecture Notes in Computer Science*, pp. 822–829.

Dayhoff, M. O., R. M. Schwartz, and B. C. Orcutt (1978). A model of evolutionary change in proteins. In M. O. Dayhoff (Ed.), *Atlas of Protein Sequence and Structure*, Volume 5(3), pp. 345–352.

Dementiev, R., J. Kärkkäinen, J. Mehnert, and P. Sanders (2008). Better external memory suffix array construction. *Journal of Experimental Algorithmics 12*, 1–24.

Dempster, A. P., N. M. Laird, and D. B. Rubin (1977). Maximum likelihood from incomplete data via the em algorithm. *Journal of the Royal Statistical Society 39*(1), 1–39.

Dohm, J. C., C. Lottaz, T. Borodina, and H. Himmelbauer (2007). SHARCGS, a fast and highly accurate short-read assembly algorithm for de novo genomic sequencing. *Genome Research 17*, 1697–1706.

Durbin, R., E. R. Sean, A. Krogh, and G. Mitchison (1999). *Biological Sequence Analysis: Probabilistic Models of Proteins and Nucleic Acids.* Cambridge: Cambridge University Press.

Dutheil, J., S. Gaillard, E. Bazin, S. Glemin, V. Ranwez, N. Galtier, and K. Belkhir (2006). Bio++: a set of C++ libraries for sequence analysis, phylogenetics, molecular evolution and population genetics. *BMC Bioinformatics 7*(188), online.

EMBL User Manual (2008). Release 96, European Bioinformatics Institute, `http://www.ebi.ac.uk/embl`.

Emde, A.-K. (2007). Progressive alignment of multiple genomic sequences. Master's thesis, Freie Universität Berlin.

Eppstein, D., Z. Galil, R. Giancarlo, and G. F. Italiano (1992). Sparse dynamic programming i: linear cost functions; ii: convex and concave cost functions. *Journal of the ACM 39*(3), 519–567.

Fabri, A., G.-J. Giezeman, L. Kettner, S. Schirra, and S. Schönherr (2000). On the design of CGAL a computational geometry algorithms library. *Software—Practice and Experience 30*(11), 1167–1202.

Ferragina, P. and G. Navarro (2008). Pizza & chili corpus, compressed indexes and their testbeds. available at `http://pizzachili.dcc.uchile.cl`, `http://pizzachili.di.unipi.it`.

Giegerich, R., S. Kurtz, and J. Stoye (1999). Efficient implementation of lazy suffix trees. In *WAE '99: Proceedings of the 3rd International Workshop on Algorithm Engineering*, pp. 30–42. Springer-Verlag.

Gotoh, O. (1982, Dec). An improved algorithm for matching biological sequences. *J. Mol. Biol. 162*(3), 705–708.

Griffith, A. (2002). *GCC: The Complete Reference.* McGraw-Hill, Inc.

Gurtovoy, A. and D. Abrahams (2002). The Boost C++ metaprogramming library.

Gusfield, D. (1997). *Algorithms on strings, trees, and sequences: computer science and computational biology.* Cambridge University Press.

Halpern, A. L., D. H. Huson, and K. Reinert (2002). Segment match refinement and applications. In *WABI '02: Proceedings of the Second International Workshop on Algorithms in Bioinformatics*, pp. 126–139. Springer-Verlag.

Henikoff, S. and J. G. Henikoff (1992, November). Amino acid substitution matrices from protein blocks. *Proceedings of the National Academy of Sciences USA 89*(22), 10915–10919.

Hirschberg, D. S. (1975, June). A linear space algorithm for computing maximal common subsequences. *ACM Press 18*(6), 341–343.

Hohl, M., S. Kurtz, and E. Ohlebusch (2002). Efficient multiple genome alignment. *Bioinformatics 33*(18), 312–320.

Holland, R., T. Down, M. Pocock, A. Prlic, D. Huen, K. James, S. Foisy, A. Dräger, A. Yates, M. Heuer, and M. Schreiber (2008). BioJava: an open-source framework for bioinformatics. *Bioinformatics 24*(18), 2096–2097.

Hopcroft, J. E. and J. D. Ullman (1990). *Introduction to Automata Theory, Languages, and Computation.* Addison-Wesley Longman Publishing Co., Inc.

Horspool, R. N. (1980). Practical fast searching in strings. *Software-Practice and Experience 10*, 501–506.

Huson, D. H., K. Reinert, S. A. Kravitz, K. A. Remington, A. L. Delcher, I. M. Dew, M. Flanigan, A. L. Halpern, Z. Lai, C. M. Mobarry, G. G. Sutton, and E. W. Myers (2001). Design of a compartmentalized shotgun assembler for the human genome. *Bioinformatics 17*, 132–139.

Hyyro, H. and H. H. Fi (2001). Explaining and extending the bit-parallel approximate string matching algorithm of myers. Technical Report A-2001-10, Department of Computer and Information Sciences, University of Tampere.

Indyk, P. and R. Motwani (1998). Approximate nearest neighbors: towards removing the curse of dimensionality. In *STOC '98: Proceedings of the thirtieth annual ACM symposium on Theory of computing*, pp. 604–613. ACM.

International Human Genome Sequencing Consortium (2001). Initial sequencing and analysis of the human genome. *Nature 409*(6822), 860–921.

ISO/IEC (1998). *Programming languages – C++, International Standard 14882* (first ed.). American National Standards Institute.

Istrail, S., G. G. Sutton, L. Florea, A. L. Halpern, C. M. Mobarry, R. Lippert, B. Walenz, H. Shatkay, I. Dew, J. R. Miller, M. J. Flanigan, N. J. Edwards, R. Bolanos, D. Fasulo, B. V. Halldorsson, S. Hannenhalli, R. Turner, S. Yooseph, F. L., D. R. Nusskern, B. C. Shue, X. H. Zheng, F. Zhong, A. L. Delcher, D. H. Huson, S. A. Kravitz, L. Mouchard, K. Reinert, K. A. Remington, A. G. Clark, M. S. Waterman, E. E. Eichler, M. D. Adams, M. W. Hunkapillar, E. W. Myers, and J. C. Venter (2004). Whole-genome shotgun assembly and comparison of human genome assemblies. *Proceedings of the national academy of science (PNAS) 101*(7), 1916–1921.

Jacobson, G. and K.-P. Vo (1992). Heaviest increasing/common subsequence problems. In *CPM '92: Proceedings of the Third Annual Symposium on Combinatorial Pattern Matching*, pp. 52–66. Springer-Verlag.

Jones, D. T., W. R. Taylor, and J. M. Thornton (1992). The rapid generation of mutation data matrices from protein sequences. *Bioinformatics, Oxford University Press 8*(3), 275–282.

Josuttis, N. M. (1999). *The C++ standard library: a tutorial and reference.* Addison-Wesley Longman Publishing Co., Inc.

Kärkkäinen, J. and P. Sanders (2003). Simple linear work suffix array construction. In *Proceedings of the 30th International Conference on Automata, Languages and Programming*, Volume 2719 of *Lecture Notes in Computer Science*, pp. 943–955. Springer-Verlag.

Kärkkäinen, J., P. Sanders, and S. Burkhardt (2006). Linear work suffix array construction. *Journal of the ACM 53*(6), 918–936.

Karlin, S. and S. F. Altschul (1990). Methods for assessing the statistical significance of molecular sequence features by using general scoring schemes. *Proceedings of the National Academy of Sciences USA 87*(6), 2264–2268.

Kasai, T., G. Lee, H. Arimura, S. Arikawa, and K. Park (2001). Linear-time longest-common-prefix computation in suffix arrays and its applications. In *CPM '01: Proceedings of the 12th Annual Symposium on Combinatorial Pattern Matching*, pp. 181–192. Springer-Verlag.

Kececioglu, J. D. (1993). The maximum weight trace problem in multiple sequence alignment. In *CPM '93: Proccedings of the 4th Annual Symposium on Combinatorial Pattern Matching*, Number 684 in Lecture Notes in Computer Science, pp. 106–119. Springer-Verlag.

Kececioglu, J. D., H.-P. Lenhof, K. Mehlhorn, P. Mutzel, K. Reinert, and M. Vingron (2000). A polyhedral approach to sequence alignment problems. *Discrete Applied Mathematics 104*(1-3), 143–186.

Kemena, C. (2008). Local and global alignment construction using the seed approach. Master's thesis, Freie Universität

Berlin, available at `http://www.seqan.de/publications/kemena08.pdf`.

Kent, W. J. (2002). BLAT – the BLAST-like alignment tool. *Genome Research 12*(4), 656–664.

Kernighan, B. W. and D. M. Ritchie (1988). *The C programming language, Second Edition*. Prentice-Hall.

Kleffe, J. and M. Borodovsky (1992). First and second moment of counts of words in random texts generated by markov chains. *Computer Applications in the Biosciences 8*(5), 433–441.

Knuth, D. E. (1997). *The Art of Computer Programming, Volume 1: Fundamental Algorithms* (3rd ed.). Addison-Wesley.

Kurtz, S., A. Phillippy, A. L. Delcher, M. Smoot, M. Shumway, C. Antonescu, and S. L. Salzberg (2004). Versatile and open software for comparing large genomes. *Genome Biology 5*(2), R12.

Lanctot, J. K., M. Li, B. Ma, S. Wang, and L. Zhang (1999). Distinguishing string selection problems. In *SODA '99: Proceedings of the tenth annual ACM-SIAM symposium on Discrete algorithms*, pp. 633–642. Society for Industrial and Applied Mathematics.

Langmead, B., C. Trapnell, M. Pop, and S. L. Salzberg (2009). Ultrafast and memory-efficient alignment of short dna sequences to the human genome. *Genome Biology 10*(3).

Larsson, N. J. and K. Sadakane (2007). Faster suffix sorting. *Theoretical Computer Science 387*(3), 258–272.

Leung, M. Y., G. M. Marsh, and T. P. Speed (1996). Over- and underrepresentation of short DNA words in herpesvirus genomes. *Journal of Computational Biology 3*(3), 345–360.

Levenshtein, V. I. (1965). Binary codes capable of correcting spurious insertions and deletions of ones. *Problems of Information Transmission 1*, 8–10.

Lim, J. H. (2007). Algorithms for motif search. Master's thesis, Freie Universität Berlin.

Lutz, M. (2006). *Programming Python.* O'Reilly Media, Inc.

Manber, U. and G. Myers (1990). Suffix arrays: a new method for on-line string searches. In *SODA '90: Proceedings of the First Annual ACM-SIAM Symposium on Discrete Algorithms*, pp. 319–327. Society for Industrial and Applied Mathematics.

Manber, U. and G. Myers (1993). Suffix arrays: A new method for on-line string searches. *SIAM Journal on Computing 22*(5), 935–948.

Manzini, G. and P. Ferragina (2004). Engineering a lightweight suffix array construction algorithm. *Algorithmica 40*(1), 33–50.

Margulies, M., M. Egholm, W. E. Altman, S. Attiya, J. S. Bader, L. A. Bemben, J. Berka, M. S. Braverman, Y.-J. Chen, Z. Chen, S. B. Dewell, L. Du, J. M. Fierro, X. V. Gomes, B.-C. Godwin, W. He, S. Helgesen, C. H. Ho, G. P. Irzyk, S. C. Jando, M. L. I. Alenquer, T. P. Jarvie, K. B. Jirage, J.-B. Kim, J. R. Knight, J. R. Lanza, J. H. Leamon, S. M. Lefkowitz, M. Lei, J. Li, K. L. Lohman, H. Lu, V. B. Makhijani, K. F. McDade, M. P. McKenna, E. W. Myers, F. Nickerson, J. R. Nobile, R. Plant, B. P. Puc, M. T. Ronan, G. T. Roth, G. J. Sarkis, J. F. Simons, J. W. Simpson, M. Srinivasan, K. R. Tartaro, A. Tomasz, K. A. Vogt, G. A. Volkmer, S. H. Wang, Y. Wang, M. P. Weiner, P. Yu, R. F. Begley, and J. M. Rothber (2005). Genome sequencing in microfabricated high-density picolitre reactors. *Nature 437*, 376–380.

Mehlhorn, K. and S. Näher (1989). LEDA: A library of efficient data types and algorithms. In A. Kreczmar and G. Mirkowska (Eds.), *MFCS*, Volume 379 of *Lecture Notes in Computer Science*, pp. 88–106. Springer.

Mehlhorn, K. and S. Näher (1999). *The LEDA Platform of Combinatorial and Geometric Computing.* Cambridge University Press.

Myers, E. W. (1999). A fast bit-vector algorithm for approximate string matching based on dynamic programming. *Journal of the ACM 46*(3), 395–415.

Myers, G. and W. Miller (1995). Chaining multiple-alignment fragments in sub-quadratic time. In *SODA '95: Proceedings of the Sixth Annual ACM-SIAM Symposium on Discrete Algorithms*, pp. 38–47. Society for Industrial and Applied Mathematics.

Myers, G. J., T. Badgett, T. M. Thomas, and C. Sandler (2004). *The Art of Software Testing*. John Wiley & Sons.

Navarro, G. (2001). A guided tour to approximate string matching. *ACM Computing Surveys 33*(1), 31–88.

Navarro, G. and R. Baeza-Yates (1999). Very fast and simple approximate string matching. *Information Processing Letters 72*(1-2), 65–70.

Navarro, G. and M. Raffinot (2002). *Flexible Pattern Matching in Strings*. Cambridge University Press.

Needleman, S. B. and C. D. Wunsch (1970). A general method applicable to the search for similarities in the amino acid sequence of two proteins. *J. Molecular Biol. 48*, 443–453.

Notredame, C., D. G. Higgins, and J. Heringa (2000). T-Coffee: A novel method for fast and accurate multiple sequence alignment. *Journal of Molecular Biology 302*, 205–217.

Pearson, W. R. (1990). Rapid and sensitive sequence comparison with FASTP and FASTA. *Methods in Enzymology 183*, 63–98.

Pearson, W. R. and D. J. Lipman (1988). Improved tools for biological sequence comparison. *Proceedings of the National Academy of Sciences 85*(8), 2444–2448.

Pitt, W. R., M. A. Williams, M. Steven, B. Sweeney, A. J. Bleasby, and D. S. Moss (2001). The Bioinformatics Template Library – generic components for biocomputing. *Bioinformatics 17*(8), 729–737.

Plauger, P. J., M. Lee, D. Musser, and A. A. Stepanov (2000). *C++ Standard Template Library*. Prentice Hall PTR.

Price, A., S. Ramabhadran, and P. A. Pevzner (2003). Finding subtle motifs by branching from sample strings. *Bioinformatics 19, supplement 2*, ii149–ii155.

Rabiner, L. R. (1989). A tutorial on Hidden Markov Models and selected applications in speech recognition. In *Proceedings of the IEEE*, pp. 257–286.

Rajasekaran, S., S. Balla, and C.-H. Huang (2005, October). Exact algorithms for planted motif problems. *Journal of Computational Biology 12*(8), 1117–1128.

Rausch, T., A.-K. Emde, D. Wecse, A. Döring, C. Notredame, and K. Reinert (2008). Segment-based multiple sequence alignment. In *Proceedings of the European Conference on Computational Biology (ECCB 2008)*.

Rausch, T., S. Koren, G. Denisov, D. Weese, A.-K. Emde, A. Döring, and K. Reinert (2009). A consistency-based consensus algorithm for de novo and reference-guided sequence assembly of short reads. *Bioinformatics 25*(9), online.

Reinert, G., S. Schbath, and M. Waterman (2005). Statistics on words with applications to biological sequences. In M. Lotaire (Ed.), *Applied Combinatorics on Words. volume 105 of Encyclopedia of Mathematics and its Applications*, pp. 252–323. Cambridge University Press.

Saitou, N. and M. Nei (1987). The neighbor-joining method: a new method, for reconstructing phylogenetic trees. *Molecular Biology and Evolution 4*, 406–425.

Sanger, F., S. Nicklen, and A. R. Coulson (1977). DNA sequencing with chain-terminating inhibitors. *Proceedings of the National Academy of Sciences 74*(12), 5463–5467.

Schulz, M. H., D. Weese, T. Rausch, A. Döring, K. Reinert, and M. Vingron (2008). Fast and adaptive variable order markov chain construction. In *Proceedings of the 8th International*

Workshop in Algorithms in Bioinformatics (WABI'08), pp. 306–317. LNBI 5251: Springer Verlag.

Schürmann, K.-B. and J. Stoye (2007). An incomplex algorithm for fast suffix array construction. *Software: Practice and Experience 37*(3), 309–329.

Sellers, P. H. (1980). The theory and computations of evolutionary distances: Pattern recognition. *Journal of Algorithms 1*, 359–373.

Shamos, M. I. and D. J. Hoey (1976, October). Geometric intersection problems. In *Proceedings of the 17th Annual Symposium on Foundations of Computer Science*, pp. 208–215.

Siek, J., L.-Q. Lee, and A. Lumsdaine (2002). *The Boost graph library: User guide and reference manual*. C++ In Depth Series. Addison-Wesley, http://www.boost.org.

Sinha, S. (2002). *Algorithms for finding regulatory motifs in DNA sequences*. Ph. D. thesis, University of Washington.

Sinha, S. and M. Tompa (2000). A statistical method for finding transcription factor binding sites. In *Proceedings of International Conference on Intelligent Systems for Molecular Biology*, Volume 8, pp. 344–354.

Sinha, S. and M. Tompa (2003). YMF: a program for discovery of novel transcription factor binding sites by statistical over-representation. *Nucleic Acids Research 31*(13), 3586–3588.

Smith, T. F. and M. S. Waterman (1981). Identification of common molecular subsequences. *Journal of Molecular Biology 147*, 195–197.

Sneath, P. H. A. and R. R. Sokal (1973). *Numerical Taxonomy: The Principles and Practice of Numerical Classification*. San Francisco: W. H. Freman.

Staden, R. (1997). A strategy of dna sequencing employing computer programs. *Nucleic Acids Research 6*(7), 2601–2610.

Stepanov, A. and M. Lee (1995). The Standard Template Library. Technical report, Hewlett-Packard Company.

Stoehr, P. J. and G. N. Cameron (1991). The EMBL data library. *Nucleic Acids Research 19*, 2227–2230.

Stroustrup, B. (2000). *The C++ Programming Language*. Addison-Wesley Longman Publishing Co., Inc.

Tarhio, J. and E. Ukkonen (1990). Boyer-moore approach to approximate string matching (extended abstract). In *SWAT '90: Proceedings of the Second Scandinavian Workshop on Algorithm Theory*, pp. 348–359. Springer-Verlag New York, Inc.

Thompson, J., P. Koehl, R. Ripp, and O. Poch (2005). Balibase 3.0: latest developments of the multiple sequence alignment benchmark. *Proteins 61*(1), 127–136.

Thompson, J. D., D. G. Higgins, and T. J. Gibson (1994). CLUSTAL W: improving the sensitivity of progressive multiple sequence alignment through sequence weighting, position specific gap penalties and weight matrix choice. *Nucleic Acids Research 22*, 4673–4680.

Tompa, M., L. Nan, T. L. Bailey, G. M. Church, B. D. Moor, E. Eskin, A. V. Favorov, M. C. Frith, Y. Fu, J. W. Kent, V. J. Makeev, A. A. Mironov, W. S. Noble, G. Pavesi, G. Pesole, M. Régnier, N. Simonis, S. Sinha, G. Thijs, J. van Helden, M. Vandenbogaert, Z. Weng, C. Workman, C. Ye, and Z. Zhu (2005). Assessing computational tools for the discovery of transcription factor binding sites. *Nature Biotechnology 23*(1), 137–144.

Ukkonen, E. (1985). Finding approximate patterns in strings. *Journal of Algorithms 6*(1), 132–137.

UniProt Consortium (2008). The universal protein resource (UniProt). *Nucleic Acids Research 39*, D190–195.

Vakatov, D., K. Siyan, J. Ostell, and editors (2003). *The NCBI C++ Toolkit [Internet]*. National Library of Medicine, National Center for Biotechnology Information, Bethesda (MD).

Vandevoorde, D. and N. M. Josuttis (2002). *C++Templates*. Addison-Wesley.

Venter, J. C., M. D. Adams, E. W. Myers, P. Li, R. J. Mural, G. G. Sutton, H. O. Smith, M. Yandell, C. A. Evans, R. A. Holt, J. D. Gocayne, P. Amanatides, R. M. Ballew, D. H. Huson, J. R. Wortman, Q. Zhang, C. Kodira, X. H. Zheng, L. Chen, M. Skupski, G. Subramanian, P. D. Thomas, J. Zhang, G. L. G. Miklos, C. Nelson, S. Broder, A. G. Clark, J. Nadeau, V. A. McKusick, N. Zinder, A. J. Levine, R. J. Roberts, M. Simon, C. Slayman, M. Hunkapiller, R. Bolanos, A. Delcher, I. Dew, D. Fasulo, M. Flanigan, L. Florea, A. Halpern, S. Hannenhalli, S. Kravitz, S. Levy, C. Mobarry, K. Reinert, K. Remington, J. Abu-Threideh, E. Beasley, K. Biddick, V. Bonazzi, R. Brandon, M. Cargill, I. Chandramouliswaran, R. Charlab, K. Chaturvedi, Z. Deng, V. D. Francesco, P. Dunn, K. Eilbeck, C. Evangelista, A. E. Gabrielian, W. Gan, W. Ge, F. Gong, Z. Gu, P. Guan, T. A. Heiman, M. E. Higgins, R.-R. Ji, Z. Ke, K. A. Ketchum, Z. Lai, Y. Lei, Z. Li, J. Li, Y. Liang, X. Lin, F. Lu, G. V. Merkulov, N. Milshina, H. M. Moore, A. K. Naik, V. A. Narayan, B. Neelam, D. Nusskern, D. B. Rusch, S. Salzberg, W. Shao, B. Shue, J. Sun, Z. Y. Wang, A. Wang, X. Wang, J. Wang, M.-H. Wei, R. Wides, C. Xiao, C. Yan, A. Yao, J. Ye, M. Zhan, W. Zhang, H. Zhang, Q. Zhao, L. Zheng, F. Zhong, W. Zhong, S. C. Zhu, S. Zhao, D. Gilbert, S. Baumhueter, G. Spier, C. Carter, A. Cravchik, T. Woodage, F. Ali, H. An, A. Awe, D. Baldwin, H. Baden, M. Barnstead, I. Barrow, K. Beeson, D. Busam, A. Carver, A. Center, M. L. Cheng, L. Curry, S. Danaher, L. Davenport, R. Desilets, S. Dietz, K. Dodson, L. Doup, S. Ferriera, N. Garg, A. Gluecksmann, B. Hart, J. Haynes, C. Haynes, C. Heiner, S. Hladun, D. Hostin, J. Houck, T. Howland, C. Ibegwam, J. Johnson, F. Kalush, L. Kline, S. Koduru, A. Love, F. Mann, D. May, S. McCawley, T. McIntosh, I. McMullen, M. Moy, L. Moy, B. Murphy, K. Nelson, C. Pfannkoch, E. Pratts, V. Puri, H. Qureshi, M. Reardon, R. Rodriguez, Y.-H. Rogers, D. Romblad, B. Ruhfel,

R. Scott, C. Sitter, M. Smallwood, E. Stewart, R. Strong, E. Suh, R. Thomas, N. N. Tint, S. Tse, C. Vech, G. Wang, J. Wetter, S. Williams, M. Williams, S. Windsor, E. Winn-Deen, K. Wolfe, J. Zaveri, K. Zaveri, J. F. Abril, R. Guigo, M. J. Campbell, K. V. Sjolander, B. Karlak, A. Kejariwal, H. Mi, B. Lazareva, T. Hatton, A. Narechania, K. Diemer, A. Muruganujan, N. Guo, S. Sato, V. Bafna, S. Istrail, R. Lippert, R. Schwartz, B. Walenz, S. Yooseph, D. Allen, A. B., J. Baxendale, L. Blick, M. Caminha, J. Carnes-Stine, P. Caulk, Y.-H. Chiang, M. Coyne, C. Dahlke, A. D. Mays, M. Dombroski, M. Donnelly, D. Ely, S. Esparham, C. Fosler, H. Gire, S. Glanowski, K. Glasser, A. Glodek, M. Gorokhov, K. Graham, B. Gropman, M. Harris, J. Heil, S. Henderson, J. Hoover, D. Jennings, C. Jordan, J. Jordan, J. Kasha, L. Kagan, C. Kraft, A. Levitsky, M. Lewis, X. Liu, J. Lopez, D. Ma, W. Majoros, J. McDaniel, S. Murphy, M. Newman, T. Nguyen, N. Nguyen, M. Nodell, S. Pan, J. Peck, W. Rowe, R. Sanders, J. Scott, M. Simpson, T. Smith, A. Sprague, T. Stockwell, R. Turner, E. Venter, M. Wang, M. Wen, D. Wu, M. Wu, A. Xia, A. Zandieh, and X. Zhu (2001). The sequence of the human genome. *Science 291*(5507), 1145–1434.

Visual C++ (2002). Microsoft Corporation. *Microsoft Visual C++.Net Language Reference*. Microsoft Press.

Wang, L. and T. Jiang (1994). On the complexity of multiple sequence alignment. *J. Comput. Biol. 1*, 337–348.

Waterman, M. S. and M. Eggert (1987). A new algorithm for best subsequence alignments with application to tRNA-rRNA comparisons. *Journal of Molecular Biology 197*(4), 723–728.

Weese, D. (2006). Entwurf und Implementierung eines generischen Substring-Index (german). Master's thesis, Humboldt-Universität zu Berlin, available at http://www.seqan.de/publications/weese06.pdf.

Weese, D. and M. H. Schulz (2008, Jul). Efficient string mining under constraints via the deferred frequency index. In

P. Perner (Ed.), *Proceedings of the 8th Industrial Conference on Data Mining (ICDM'08)*, pp. 374–388. LNAI 5077: Springer Verlag.

Weiner, P. (1973). Linear pattern matching algorithms. *Annual Symposium on Switching and Automata Theory 0*, 1–11.

Wilson, M. (2004). *Imperfect C++: Practical Solutions for Real-Life Programming*. Addison-Wesley Professional.

Wöhrle, H. (2006). Multidimensionales Chaining mit Deferred Range Trees (german). Master's thesis, Freie Universität Berlin, available at `http://www.seqan.de/publications/woehrle06.pdf`.

Wu, S. and U. Manber (1992). Fast text searching: allowing errors. *Communications of the ACM 35*(10), 83–91.

Wu, S. and U. Manber (1994). A fast algorithm for multi-pattern searching. Report TR-94-17, Department of Computer Science, University of Arizona.

Zerbino, D. R. and E. Birney (2008). Velvet: Algorithms for de novo short read assembly using de bruijn graphs. *Genome Research 18*, 821–829.

Zhang, Z., S. Schwartz, L. Wagner, and W. Miller (2000, Feb). A greedy algorithm for aligning dna sequences. *Journal of Computational Biology 7*(1/2), 203–214.

Index

Printed and bound by CPI Group (UK) Ltd, Croydon, CR0 4YY

26/10/2024

01779749-0002